ALIENS

A L I E N S

P A S T   P R E S E N T   F U T U R E

R O N   M I L L E R

Forewords by DAVID BRIN and DR JOHN ELLIOTT

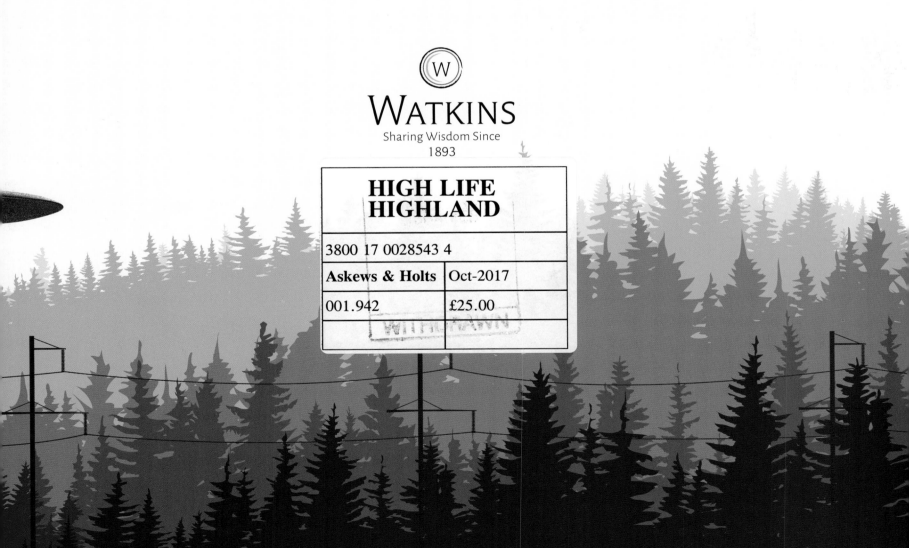

W

WATKINS
Sharing Wisdom Since
1893

© 2017 Elephant Book Company Limited

This 2017 edition published by Watkins by arrangement
with Elephant Book Company Limited,
Southbank House, Black Prince Road,
London SE1 7SJ, United Kingdom

Watkins, an imprint of Watkins Media Limited
19 Cecil Court,
London WC2N 4EZ, United Kingdom

enquiries@watkinspublishing.com

1 3 5 7 9 10 8 6 4 2

Editorial Director: Will Steeds

Project Editor: Chris McNab

Designer: Paul Palmer-Edwards, Grade Design

Picture Researcher: Susannah Jayes

Colour reproduction: Pixel Colour Imaging

Printed and bound in China

A CIP record for this book is available from the
British Library.

ISBN: 978-1-78028-968-7

www.watkinspublishing.com

For permission to reproduce illustrations appearing in
this book, please correspond directly with the owners
of the works, listed on p.224. Elephant Books does not
retain reproduction rights for these images individually,
or maintain a file of addresses for sources.

The information in this book is true and complete
to the best of our knowledge. All recommendations are
made without any guarantee on the part of the author
or publisher, who also disclaim any liability in connection
with the use of this data of specific details. We recognize,
further, that some words, model names, and designations
mentioned herein are the property of the trademark
holder. We use them for identification purposes only.
This is not an official publication. While every attempt
has been made to establish copyright for the images
reproduced in this book, this has proved impossible in
a few cases. Elephant Book Company Limited apologizes
for any inadvertent infringement of copyright, and will
be grateful for notification of any errors or omissions.

Front cover illustration: Paul Palmer-Edwards

CONTENTS

FOREWORD #1
DAVID BRIN – ON CONTEMPLATING THE ALIEN

There is a glass ceiling. Pressing just below it, a range of candidates jostle and chatter. Dolphins and apes. Elephants and sea lions. Parrots and crows. Prairie dogs and even octopuses. Every few years, scientists discover yet another species sharing at least a little of what we thought was unique to us. Some degree of "language" communication skill. Some ability to use tools.

To many people, these revelations constitute a breakthrough in maturity, as humanity recognizes that we aren't the only beings with inner lives, or with some facility in logic and conversation. There is a kind of humbling pleasure in realizing that "we're not so special, after all." And yet, look again at that clustering. So many species with very basic abilities to "speak" and manipulate, but none of them rising above that rigid ceiling; it is as if both Nature and Darwin are saying: "Thus far, many may rise easily. Beyond, though, will be hard." Does the clustering actually say that *Homo sapiens* are very special? Humans, for well or ill, smashed through successive glass ceilings, starting a million years ago and, according to evidence, many times since.

All creatures live embedded in time, though we are self-conscious enough to comment on it, lamenting the past or worrying over the future. Obsession with either past or future can define a civilization. Worldwide, most historical cultures believed in some lost Golden Age when people knew more, mused loftier thoughts and were closer to the gods – but then fell from grace. Under this dour but recurrent world view, men and women of a later, coarser era can only look back with envy, harkening back to remnants of ancient wisdom. Just a few societies dared contradict this standard dogma of nostalgia. Our own scientific West, with its impudent notion of progress, brashly relocated any Golden Age to the future, something to work toward, a human construct for our grandchildren to achieve with craft, sweat and goodwill – assuming that we manage to prepare them.

And not only forward in time. We are also beings who look outward, seeking different faces, voices. Elsewhere – in a book called *Otherness* – I talk about how this itch to make contact may be rooted in our nature, as gregarious apes, but has amplified greatly, in this new kind of brash and yearning society. It's one reason why we love contemplating the alien. That itch and search – portrayed so well by Ron Miller in this book – was pondered by a few even thousands of years ago. But the interest became an obsession that has strengthened across the last century, manifesting in our space programmes, in serious searches for extraterrestrial life like SETI (Search for Extraterrestrial Intelligence), and especially in science fiction.

Which brings us to this volume, a feast for both the curious mind and your aesthetic eye. Ron Miller's art has always conveyed unrestrained verve, creativity and courage, as well as joyful curiosity at the endless possibilities of our cosmos. Now – in *Aliens: Past Present Future* – he takes us on a guided tour of otherness, illustrating how our concept of extraterrestrial life has evolved across the centuries. And what a tour! Come along and have fun. Be inspired by strange extrapolations of what might lie just beyond the curtain . . . or our own glass ceiling. And know that what we'll find, someday, may be even stranger still.

OPPOSITE: An illustration from *Entretiens sur la Pluralité des Mondes* (Conversations on the Plurality of Worlds; 1686), in which Bernard Le Bovier de Fontenelle made one of the first attempts to describe the inhabitants of other worlds based on the physical conditions of the planet on which they evolved.

FOREWORD#2
DR JOHN ELLIOTT – ON ALIEN COMMUNICATION

The core beat of life, once established, is to survive and thrive in its environment. Our planet Earth constantly reveals new examples of this adaptability, especially in its most extreme "corners", where the environmental conditions shape living creatures that are almost alien to our own, human solution.

Whether the eventual encounter with intelligent life from another world results in our making contact with physiologies close to our own (as often historically predicted or imagined), or solutions beyond our conventional expectations (whatever their scale), it is unlikely to exceed the realms of our imagination. What is true, however, is that the most unlikely scenario is that we are alone in this vast Universe.

As our technology and understanding advances, our ability to study and reveal once-hidden planets around distant stars is ever increasing and many other worlds are now in evidence, even in our own galactic "backyard". If there is one observable attribute that clearly demonstrates an advanced intelligence, it is the ability to communicate knowledge asynchronously: across both space and time. On our own planet, Earth, we demonstrate the facility to retain knowledge and thereby build upon our collective wisdom. For example, my contribution, since the late 1990s – which quickly resulted in working with the SETI (Search for Extraterrestrial Intelligence) Institute – has focused on how we might detect and understand such communication from an extraterrestrial being and discover what is special about language in the "signal" universe that might help us recognize it.

Prior to this, most scientific endeavours were focused only on detecting the existence of an extraterrestrial technology. In this new pursuit, the study of language types across human history has given insight into the core constructs of communication, beyond the apparently arbitrary pairing of meaning with symbols and sounds. Here we have found the "fingerprint" of our cognition and *Lingua Ex Machina* (brain) and thereby the beginnings of a universal template for recognizing as yet unknown systems carrying information – messages, in short. Non-human examples of communication, such as those expressed by the dolphin, ape, bird and robot, have also added to this template.

Like any venture into the unknown, we have to begin somewhere, and from a scientist's point of view the best foundation is first to look at what we already know, before expanding our horizons beyond this frontier. As with the environmental drivers that mould physical variations, tribalism has been the engine of our language diversity. For other civilizations, however, this aspect, born out of community and place, may have eroded. Our physiology directly impacts on our use of communication, a survival-critical attribute; so, although science rationalizes and provides sound methods for deriving evidence-based knowledge, it is sometimes an actual "leap of faith" – imagination – that can provide the bridge to an eventual discovery.

We must also accept that there are many examples of "truths" and "absolutes" being proved inaccurate or wrong. We once believed the world to be flat and at the centre of the universe, for example, so we should always retain a healthy scepticism toward today's absolutes. The list of things we still do not know is a very long one indeed. Our mantra should be to keep

RIGHT: The Hubble Telescope here captures a towering pillar of gas and dust, three light years tall, in the Carina Nebula. New technology is constantly driving our vision deeper into the universe, increasing the possibilities of the detection of extraterrestrial life.

open minds and embrace new thinking as, all too often, conventional wisdom has taken us down the wrong path. It is also worth remembering that other intelligent "life" forms may be post-biological and, therefore, not subject to the constraints we attribute to biological solutions.

Nevertheless, when contact is made, we will need to be as prepared as we possibly can be. In this related endeavour, and informed by work on communication, our protocols and strategies for post-detection scenarios and message response construction form important additional and complementary pursuits. The academic community (via the International Astronautics Association's SETI Committee) has established an agreed voluntary protocol to guide our actions when contact is made. Outside academia, however, there is no working international protocol as yet and therefore no advanced strategy agreed by the world's governments. At best, this is a work

in progress. On a positive note, contact will most likely be via a "signal" from a distant location, which will at least allow us some time (measured in years) to consider our options and analyze the phenomenon.

In this book, aided by a wealth of imagery, we are taken on a quest through our imaginations, postulations, speculations and past claims, to help us ponder the "Alien" question, but we are always balanced and grounded by the author's reflections. This journey takes us from our early discoveries about the universe and ideas on how extraterrestrial life may present itself, including through popular culture and myth, to some of our current and possible future scientific endeavours. In doing so, it provides a refreshingly objective look at our visions of extraterrestrial life, reflecting on one of the most fundamental questions contemplated by the human race: Are we alone?

INTRODUCTION

"Empty space is like a kingdom, and earth and sky are no more than a single individual person in that kingdom. Upon one tree are many fruits, and in one kingdom there are many people. How unreasonable it would be to suppose that, besides the earth and the sky which we can see, there are no other skies and no other earths."

Teng Mu, a Chinese scholar of the Sung Dynasty (AD 960–1280)

It's fun to speculate about what life might be like on other worlds. What kinds of weird creatures might live on Mars? Pluto? Planets orbiting other stars? It's not only a great question to ask, but it's an important question, too. For one thing, to answer it we have to look at how life came to be on our own planet. Why do the things that live on the earth look the way they do? How much is due to the world we live on, and how much is due to pure chance? However and wherever life might exist, does it inevitably lead to intelligent beings that resemble us? Or could it lead down paths we can scarcely imagine?

Perhaps even bigger questions relate to our place in the universe. Are we alone? Or do we share the cosmos with countless other species? Is it possible that of the one billion trillion stars in the observable universe, one and only one has a planet on which intelligent life evolved? These cannot be considered trivial questions.

So, are we indeed alone in the universe? Is ours the only solar system? Is the earth the only planet with life? As the 21st century opens, it appears that astronomers are about to provide an answer, with profound possibilities. As the great scientist and author Arthur C Clarke observed, "Sometimes I think we're alone in the universe, and sometimes I think we're not. In either case the idea is quite staggering."

Speculation about life on other worlds has had an immeasurable impact on our culture as well. Much of this impact has been for amusement, generating tens of thousands of science fiction stories, novels, comic books and movies. But even entire philosophies and religions have been based on the idea that there is life on other planets, revealed in messages to select initiates by alien visitors. Many of these beliefs are perfectly sincere … others perhaps not so much so.

Some of these beliefs may in fact be the last remnants of the geocentrism that drove astronomy for the thousands of years that preceded Copernicus. It is part of the conceit that is hardwired into every human to want our planet – and our species in particular – to be not

COLLIER'S   GLENN R. BERNHARDT

ABOVE: Long before the idea of extraterrestrial life became part of our popular culture, let alone of interest to scientists, science fiction illustrators, such as Leo Morey, delighted in depicting the bizarre life forms that might exist on other worlds, as shown here in his drawing from 1935.

ABOVE RIGHT: The flying saucer craze of the 1950s was parodied in this cartoon by Glenn R Bernhardt published in *Collier's* magazine. Pairing a visiting alien with children in spacemen costumes also played off the popularity of the many science fiction television programmes of the day, such as *Space Patrol*.

only the centre of the universe, but also the most important thing in the universe. The idea itself is central to most religions, which often believe that not only does the entire universe whirl around our heads, but was expressly *created* for us. So it's perfectly natural for humans to assume that if there are other intelligent beings in the cosmos, we would be the centre of their attention.

It might be worthwhile to mention at this point what this book is *not* about, although this theme is touched on in parts. And that is the subject of UFOs, with which there are two problems. The first is that there are any number of theories as to just what UFOs might actually be. Probably the most popular is the Extraterrestrial Hypothesis: that UFOs are spacecraft piloted by beings from other planets. The difficulty lies in there being many competing theories, some of which I will be discussing and all of which are equally valid. The subject of flying saucers and UFOs is certainly part of the story of aliens

and extraterrestrial life and cannot be entirely ignored, at least so far as its impact on the popular conception of extraterrestrials. On the other hand, we want to focus on how life might have evolved on other worlds and what that life might be like. The second problem is that, as interesting as UFOs might be, spending too much time discussing them would be like focusing on the *Spirit of St Louis* or Apollo 11 instead of pilot Charles Lindbergh or astronauts Armstrong, Aldrin and Collins. Even if UFOs do turn out to be piloted by aliens from other worlds, we should be talking about who or what the pilots are, and concern ourselves with how they got here. While certainly germane to a degree in any discussion about alien life, UFOs – whether they even exist at all – are really an entirely different subject (to which entire libraries of books have been devoted) and we need to be cautious about getting too far afield from the real subject of this book: the aliens themselves.

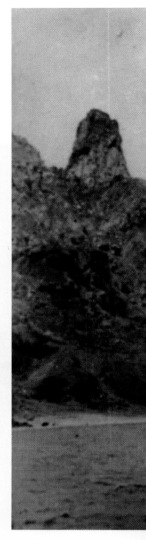

OPPOSITE, TOP: What appears to be an alien invasion in the 1952 Austrian film *1 April 2000* proves to be in reality a spaceship sent by a future World Court to pass judgment on the country. Like many other filmmakers at the time who featured spacecraft in their work, director Wolfgang Liebeneiner made his seem more "real" by following the pattern described by numerous UFO observers.

OPPOSITE, BOTTOM LEFT TO RIGHT: For nearly seventy years, *Fate* magazine has reported on flying saucer sightings, alien encounters and the many other manifestations of the public's interest in extraterrestrial life and the possibility of visitors to Earth from other worlds.

ABOVE: This famous photograph of a UFO was taken in 1958 by Almiro Baraúna on the Brazilian island of Trinidade. It was eventually revealed in 2010 to have been a hoax.

But even if we give only passing acknowledgement to the question of UFOs and flying saucers, that still leaves an enormous subject, one that can only be introduced in a book as relatively brief as this one. Until recently, the only example we had of life in the entire universe were the lifeforms we see on our own planet. Even today, we know of nowhere else among billions of galaxies and trillions of stars where life exists. Yet … there are tantalizing clues. We now know that the basic building blocks of life – organic molecules and compounds, amino acids and water – are found in abundance everywhere we look. Within our own solar system, we know that the planet Mars was once rich with water – and may still be so today, even if it is in the form of ice buried far underground. Saturn's giant moon, Titan – a world large enough to be a planet in its own right were it circling the Sun directly – has a surface covered in organic material. Oceans of water lie beneath the icy crusts of Europa and Enceladus, rich in organic compounds. And in recent years some astronomers have discovered planets beyond our solar system that may be amenable to the evolution of life. The star nearest our own solar system – Proxima Centauri, only 4.5 light years away from us – has a planet nearly the same size as Earth. It orbits within Proxima Centauri's "Goldilocks zone" – that region around a star which is neither too hot nor too cold to support life – and may possess liquid water on its surface.

So the answer to Clarke's question may well turn out to be that we are not only not alone, but live in a universe teeming with life.

01

"Do there exist many worlds, or is there but a single world? This is one of the most noble and exalted questions in the study of Nature."

Albertus Magnus, c13th century

WORLDS OTHER THAN OUR OWN

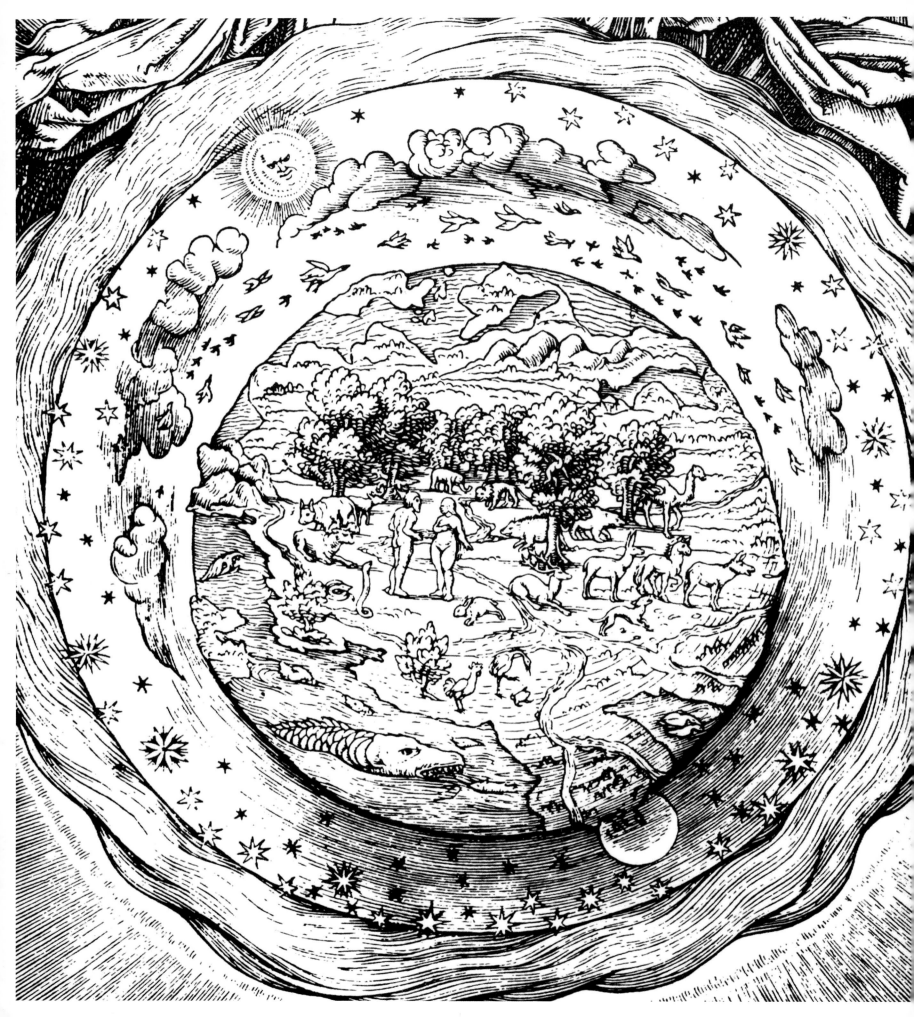

THE DISCOVERY OF THE UNIVERSE

Before the night of 6 January 1610, most human beings were certain that the world as they knew it occupied a special place, a unique position, in the cosmos. Surely anyone could see with their own eyes that the Sun, the Moon and the stars circled Earth. Moreover, there were no worlds other than this one. Common sense told them these things, and so did their religions. Judeo-Christian tradition – based on an even earlier Babylonian cosmology – insisted that Earth was flat and unmoving, that the Sun and Moon went around it and that the sky was a solid dome overhead. These ideas did not begin to change until a handful of philosophers started to ask questions about the universe they saw around them. For one thing, they started wondering if Earth circled the Sun instead of the other way around. The theory was difficult to prove and it went against both common sense and simple observation: it was obvious to everyone that the Sun went around Earth just like the stars did and that Earth was, consequently, in the middle of the entire universe where it belonged.

OPPOSITE: This illustration from Martin Luther's *Biblia* (1534) depicts Earth with the Garden of Eden at its centre, surrounded by a firmament (or sky) containing the Sun, Moon and stars.

RIGHT: This is a road map of our solar system. Reading outward from the Sun (starting with the inset) are the orbits of Mercury, Venus, Earth and Mars (red), followed in the main diagram by Jupiter, Saturn, Uranus, Neptune and Pluto (white).

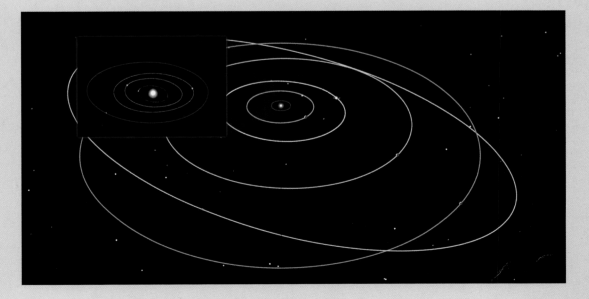

> "For everyone . . . must see that astronomy compels the soul to look upward"
>
> Glaucon in Plato's *The Republic* (380 BC)

ABOVE: Earth as described in the Bible owed a lot to earlier Babylonian concepts. Earth was flat and supported by pillars. Beneath Earth were fountains of water and the underworld, above was the solid dome of the heavens.

A New Look at the Universe

New scientific ideas about the universe were not particularly popular. The Greek scientist and philosopher Anaxagoras was banished from Athens around the year 434 BC for suggesting that the Sun was a white-hot stone no larger than the Peloponnese, a peninsula in southern Greece covering about 1,000 square miles. But while ancient scientists were attempting to discover the true size and distance of the Sun, only a very few ever questioned the "fact" that the Sun went around the Earth – and those few who suggested otherwise were never taken *really* seriously. It seemed to be just plain common sense – it was something anyone could see happen every day. Surely, they stated, Earth was unmoving as the Sun rose in the east and set in the west. This also fitted in with religious ideas about the place Earth held in the universe: at the very centre, with everything else, the Sun, the Moon, the stars and the planets, orbiting in circles around it. To doubt this seemed not only nonsensical but possibly even heretical.

This situation did not begin to change until the Greek scientist Eudoxus started to think about what might really be happening in the sky. This was around the year 400 BC. Like most people at the time, he believed that Earth lay at the centre of the universe. But this did not explain some of the motions he observed. For example, there was the problem presented by five special stars. All the other stars in the sky moved together from east to west. Because they did not seem to move in relation to one another they were called the fixed stars. But there was a small group of stars that moved among the fixed stars. The Greeks called these odd stars the *wanderers* and in Greek the word for *wanderer* was *planete*.

To explain this motion, Eudoxus suggested that Earth lay motionless at the very centre of a series of transparent spheres. Each of these spheres was free to rotate in any direction.

ALIENS ON EARTH:
Speculation about Life in Strange, Distant Lands

Many hundreds of years ago, most of our own planet was as unknown as the far side of the Moon once was. North and South America, the depths of the African continent, the faraway lands of Asia were all as alien to early European explorers as Mars, Venus or Jupiter are today. Even more so: we could at least *see* those worlds far off in the sky. After the 17th century anyone with a telescope could explore the Moon, but in order to explore the unknown lands being discovered across the seas one had to actually *go* there. In fact, detailed maps of Earth's Moon and Mars were created long before the interior

lands of the New World were charted. As explorers began to discover and investigate the new-found worlds that lay on the far side of the oceans or deep within impenetrable continents, they brought back reports of the astonishing things they witnessed. In some cases, they simply didn't understand what they saw. Weird native costumes, headdresses and body paint may have misled a startled European explorer into thinking he'd seen something inhuman and monstrous. Or someone may have jumped to a conclusion based on a mistranslation or misunderstanding. Strange

ABOVE: This 16th-century woodcut by Hans Weiditz II illustrates what might be a modern encounter with an extraterrestrial visitor.

animals, such as giraffes, rhinoceros or elephants may have stretched an observer's descriptive abilities to their limit. In other cases, an explorer may have invented weird creatures out of whole cloth in order to make their adventures sound more exciting. Map-makers and illustrators, too, often let their imaginations run wild when descriptions seemed vague or odd or there were blank spaces in a chart to fill.

"...my own suspicion is that the Universe is not only queerer than we suppose, but queerer than we can suppose."

J B S Haldane (1927)

The fixed stars were attached to the outer sphere and their slow motion through the night sky was caused by the rotation of the sphere. The next largest sphere carried the *wanderer* called Saturn. Jupiter, Mars, the Sun, Venus, Mercury and the Moon were each attached to the remaining spheres. Since these spheres could move independently, the planets appeared to move against the background of fixed stars. Most of the observed motions were explained by this elegant but complicated system, but there were still a lot of questions unanswered.

Aristotle, another Greek scientist, had no doubts about Earth's place at the centre of everything. He accepted Eudoxus' concentric spheres, but also reintroduced the idea of Earth itself being spherical. After all, he asked, you can see the shadow of Earth on the Moon during a lunar eclipse. The shadow has a curved edge so he thought it was reasonable to suggest that Earth itself must be round.

Aristarchus, a Greek scientist who lived from 310 BC to *c*230 BC, not only believed that Earth was a sphere, but he may also have been the first scientist to suggest that Earth circled the Sun just as the stars and planets did. The appearance of the stars circling Earth, he said, was only an illusion. They seemed to move through the sky because Earth itself was rotating on its axis. Although he was perfectly correct, his ideas were considered to be little more than an interesting philosophical thought experiment and they were soon forgotten.

More than 300 years later, the Greek scientist Claudius Ptolemaeus (usually known as Ptolemy) published an explanation for the universe that was to influence thinking for the next 1,500 years.

He was strongly in favour of a non-rotating Earth that lay at the centre of the universe. Ptolemy accepted Eudoxus' crystal spheres to explain the motions of the Moon, stars and planets but realized that they did not explain every observed movement of the planets. He refined the theory by adding a complicated arrangement of overlapping circles. Each of the five known planets moved in a perfect circle around Earth but to explain the observed deviations he proposed that as each planet moved along its orbit, it also moved in a smaller circle called an *epicycle*. Although Ptolemy's system required a kind of complex clockwork of overlapping circles, it *did* explain the movement of the stars and planets while at the same time keeping the Earth motionless at the centre. And because that fitted neatly within the rapidly growing and increasingly influential faith called Christianity, it became dogma that could not be questioned. In fact, it became dangerous to challenge it.

Galileo's Great Discovery

Up until just over 400 years ago, people considered the planets interesting but of no special significance; the only thing that set them apart from the thousands of other stars was that they moved. No one ever imagined that they might in fact be other *worlds*. Not until the year 1610, that is.

The Italian scientist Galileo Galilei (1564–1642) had been experimenting with an amazing new device recently introduced from the Netherlands. Consisting of nothing more than a pair of ordinary magnifying lenses set at either end of a wooden or cardboard tube, it had the remarkable property of making distant objects appear closer. Others

ABOVE: Galileo Galilei (1564–1642) was the first to turn a telescope toward the night sky. What he discovered when he looked heavenwards changed forever humanity's perception of its place in the universe.

ABOVE LEFT: William Herschel (1738–1822) was a professional musician who gained his greatest fame as an astronomer.

ABOVE RIGHT: John Couch Adams (1819–92) was a British astronomer who proved the existence of Neptune using only mathematics.

were already aware of its potential use to navigators and the military, but Galileo did something with the telescope that no one had ever considered: he turned it toward the night sky.

What Galileo learned forever changed how we regard the universe around us and even how we regard Earth itself and its place in that universe. He saw that the Moon was "not smooth, uniform and precisely spherical as a great number of philosophers believe it (and the other heavenly bodies) to be, but is uneven, rough and full of cavities and prominences, being not unlike the face of the Earth" (*Sidereus Nuncius* / "The Starry Messenger", 1610). The planets, he found, were not just a special class of star, but were in

fact worlds, perhaps very much like our own. They were spherical, like Earth, and some of them, like Mars, had vague markings that might, Galileo thought, be continents and seas. Venus and Mercury showed phases like the Moon – proving that they circled the Sun – but Saturn was a real mystery. It appeared to have weird appendages like ears or handles (it took 40 years and an astronomer with a better telescope to discover that the "ears" were in fact a set of glorious rings circling the planet). Galileo was even more astonished to find that Jupiter was not only a world, but that it had four moons of its own (though now we know that it possesses at least 67!). It was like a miniature solar system in its

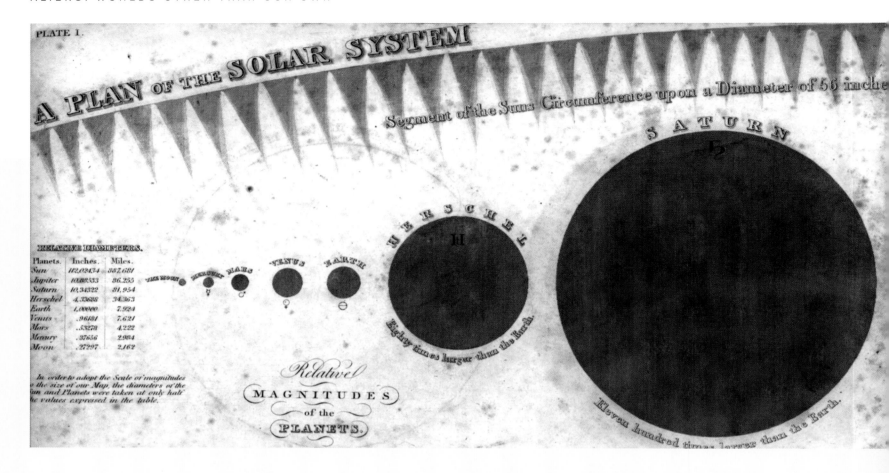

PLATE 1.

A PLAN OF THE SOLAR SYSTEM

Segment of the Suns Circumference upon a Diameter of 56 inches

SATURN

HERSCHEL

Eighty times larger than the Earth.

Eleven hundred times larger than the Earth.

RELATIVE DIAMETERS.

Planets.	Inches.	Miles.
Sun	112,02434	887,681
Jupiter	10,80333	86,255
Saturn	10,34322	81,954
Herschel	4,33688	34,363
Earth	1,00000	7,924
Venus	,96181	7,621
Mars	,53278	4,222
Mercury	,37656	2,984
Moon	,27297	2,162

In order to adopt the Scale or magnitudes to the size of our Map, the diameters of the Sun and Planets were taken at only half the values expressed in the table.

Relative
MAGNITUDES
of the
PLANETS.

THE MOON   MERCURY   MARS   VENUS   EARTH

More Planets?

While surveying the stars one night in 1781, self-taught British astronomer William Herschel discovered an entirely new planet. Until then, astronomers had not bothered to look for other planets because they assumed there *couldn't* be any more than the five planets already known – Mercury, Venus, Mars, Jupiter and Saturn. Together with the Sun and Earth this made seven celestial bodies. Many people – especially Christians – considered this number important. Since God had created the universe in seven days, seven came to be associated with completeness and perfection.

At first Herschel thought he had found a new comet, but comets have highly elliptical orbits that are vastly different from the nearly circular orbits of most planets. Uranus, as the new planet

was eventually named (after Urania, the muse of astronomy), can barely be detected by the naked eye. It is so faint and so distant that its motion among the stars is scarcely noticeable. In fact, it has only circled the Sun about two and a half times since it was discovered.

Astronomers eventually noticed something odd about the new planet. Uranus didn't seem to move as calculations predicted it should. In some years it seemed to lag behind, and in other years it seemed to move too quickly. In 1834 Reverend T J Hussey of Kent, England made a startling suggestion: what if there was yet another unknown planet orbiting beyond Uranus? The gravitational pull of the mystery planet upon Uranus might account for the inaccuracies. When it was ahead of Uranus, it tugged that planet a little farther ahead than it should be. When it lagged behind Uranus, its pull slowed it down.

Hussey thought it might be possible to predict the location of this mysterious planet by studying its effect on Uranus. By 1843 John Couch Adams,

ABOVE: This chart of the planets published in 1835 includes the planet "Herschel" (second from the right), named in honour of its discoverer. This was soon changed to "Uranus", the name by which we know it today.

" . . . we may have an idea of the numberless globes that serve for the habitation of living creatures."

William Herschel (1795)

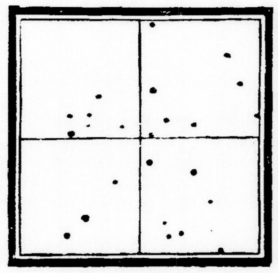 

ABOVE: Johann Galle's (1812–1910) discovery drawing, indicating with an arrow the position of the new planet, Neptune, among the known fixed stars. The small white cross shows where Urbain Le Verrier (1811–77) had predicted that the planet would be found, less than a degree away.

a student at the University of Cambridge, had calculated where he thought the new planet should be found. He sent his results to George Airy, the British Astronomer Royal, but Airy did nothing with Adams' calculations until 1846. By that time the French astronomer Urbain Le Verrier had published the results of his own independent calculations. Le Verrier's predicted location for the new planet was almost exactly the same as Adams'. However, unlike Adams' work, Le Verrier's was immediately published. As soon as Airy learned of this event, he assigned two astronomers – James Challis and William Lassell – to search for the unidentified planet. If a Frenchman had been first to announce the location of a new planet, Airy was determined that it would be a British astronomer who first actually saw it.

Challis spotted what he thought was the new planet on 4 August and again on 12 August, but failed to check his observations. However, in the meantime, Johann Galle and Heinrich d'Arrest,

both of the Berlin Observatory in Germany, found and identified the planet, which was named Neptune because its greenish colour recalled the Roman god of the sea.

Astronomers became confident that what had been done once might be done again. John Pringle Nichol in *The Planet Neptune* (1848) quotes Le Verrier as saying: "This success permits us to hope that after thirty or forty years of observation on the new Planet [Neptune], we may employ it, in its turn, for the discovery of the one following it in its order of distances from the Sun."

A False Start

After the discovery of Neptune, astronomers were eager to see if other new planets lurked on the frontiers of the solar system. While some searched the void beyond Neptune, others wondered if a planet might be orbiting the Sun closer than Mercury. It had long been observed that there were unexplained disturbances to Mercury's orbit. These were similar to the

"That planet has a considerable but moderate atmosphere."

William Herschel (1784)

OPPOSITE: When Mercury transits (passes in front of) the Sun, it appears as little more than a miniscule black dot against the incandescent surface of the star.

disturbances to Uranus' orbit that led to the discovery of Neptune. It was assumed that the disturbances were caused by an unknown planet.

The problem with trying to observe a planet close to the Sun is caused by the Sun itself – its glare makes it almost impossible to see anything near it. Mercury is difficult enough to observe. Anything closer to the Sun is worse. But every now and then, the orbits of Mercury and Venus cause them to pass in front of the Sun as seen from Earth. When this happens, the planets appear like tiny round black dots silhouetted against the bright face of the Sun. Using telescopes especially equipped to observe the Sun, astronomers started searching the face of it, looking for an unfamiliar small black dot – a job that was made especially difficult since the Sun is already freckled with numerous sun spots, any

one of which might be mistaken for the mysterious planet.

Astronomers searched for nearly half a century without finding anything until 1859, when French astronomer Edmond Lescarbault announced that he had discovered the new planet. Claiming that he had observed a small round black spot cross the face of the Sun at a time when neither Mercury nor Venus would have been visible, he convinced several leading astronomers, including Urbain Le Verrier (still famous from his co-discovery of Neptune), of the reality of his find. The new planet was named Vulcan, after the Roman god of fire. Le Verrier – perhaps overconfident after his success with Neptune – predicted that it would reappear on 29 March, 2 April and 7 April 1860. Hundreds of astronomers waited hopefully for Vulcan's appearance on those dates, but in vain. The new planet failed to appear. In fact, no one has ever seen it again.

Eventually, the mysterious aberrations in Mercury's orbit were accounted for by Einstein's "General Theory of Relativity", which explained how space time is disturbed by the Sun's mass. Interestingly, Mercury's orbit was one of the tests that helped establish the theory's validity.

LEFT: William Herschel's 50-inch (1.26m) telescope was built between 1785 and 1789 and was the largest telescope in the world until the 72-inch (1.83m) "Leviathan of Parsonstown" was built in 1845.

"Bright points in the sky or a blow on the head will equally cause one to see stars."

Percival Lowell (1895)

New Worlds around the Sun

After this landmark theoretical breakthrough in the history of science, the search for a planet beyond Neptune was long considered a hopeless quest. The method used in the 19th century to discover Neptune becomes more difficult the farther a planet is away from the Sun. This is because the planets that are farthest away move much slower in their orbits. With a Neptunian year being 165 Earth years long, it would take an extremely long time for Neptune to move far enough for even the sharpest-eyed astronomer to be able to detect disturbances that might be caused by an even more distant world. Indeed, the planet has barely completed even one Neptunian year since its discovery. Yet in the early 20th century a wealthy amateur astronomer from Boston took up the challenge to find a planet beyond Neptune.

Percival Lowell, a brilliant mathematician, had graduated from Harvard in 1876 with honours. He had long had an interest in astronomy and decided to make it his career, specializing in the planet Mars. With his wealth he built and funded his own observatory near Flagstaff, Arizona on a 2,133m (7,000ft) elevation he called "Mars Hill". (The observatory he founded continues in operation to this day.)

Although his main interest lay in Mars (see Chapter 4), Lowell eventually began to wonder if it was possible to determine the existence of a new outer planet by ignoring Neptune and focusing instead on Uranus. To do so, he would have to calculate the planet's orbit with a great deal more accuracy than either Adams or Le Verrier had done. His reasoning was that any disturbances in Uranus, however slight, that

Neptune could not account for must be caused by yet another planet. These calculations consumed many years, but in 1905 Lowell announced that he had determined the orbit of Planet X. He believed it was a small world 4 billion miles from the Sun – more than 40 times farther than Earth – taking 282 Earth years to make a single orbit. Something so small and so far away would be almost impossibly faint and consequently even more difficult to discover than Neptune had been.

However, Lowell had a significant advantage over Adams and Le Verrier, who had to do their observations visually and so could only be as good as their eyes and their hand-drawn charts could make them, whereas Percival Lowell had cameras. He simply had to take a photograph each night of one small part of the sky at a time. By comparing photographs, he could tell if any one of the tens of thousands of points of light had moved.

But even on a photographic plate, the object for which Lowell searched would be extremely small and dim. And there are many other small, dim objects that move in the night sky, such as comets and asteroids. False leads like these had to be laboriously eliminated. By the time Lowell died in 1916, Planet X remained undiscovered.

Other astronomers tried their luck, but they met no better success, and interest in the new planet slowly died out. A new telescope was installed at Lowell Observatory in 1929, bringing fresh hope to the search. The camera telescope with its 33cm (13in) lens could detect objects many times dimmer than the instrument Lowell had been using. A 23-year-old astronomer named Clyde Tombaugh was assigned to the search for

TOP: Dilettante astronomer Percival Lowell (1855–1916) financed and encouraged the search for "Planet X".

BOTTOM: A young Clyde Tombaugh was assigned the task of examining hundreds of photographic plates, looking for the elusive speck that was the new planet.

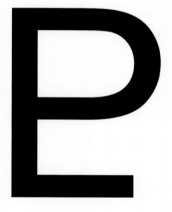

TOP LEFT & RIGHT: When Tombaugh saw that one of the "stars" (indicated by the arrows) appeared to jump back and forth as these two plates were compared, he realized that he had found the sought after "new world".

BOTTOM LEFT: The astronomical symbol for the new planet combined the first two letters of its name, "Pluto". It was also, purposefully, the initials of Percival Lowell, who had for so long championed its discovery.

Planet X. Even with the new equipment, Tombaugh's task was daunting. The camera photographed only a tiny portion of the night sky. Two or three days later, the same location would be photographed again. Tombaugh would place the two photographs into a device called a blink comparator. This allowed him to see first one and then the other photo. He compared them by flipping the images back and forth quickly, like the frames in an animated cartoon. Fixed points of light, such as stars, remained unchanged. But if something had moved between the two exposures, it appeared to jump back and forth. Since each image contained as many as 160,000 tiny points of light – or 400,000 if any part of the Milky Way was in the plate – the project took an entire year to complete. As a precaution, Tombaugh took three plates of each star field. This extra plate acted as check against misidentifying a flaw in one of the plates as the object of his search. This added an additional two to three hours of work every day; he was already spending six or seven hard hours sat at the blink comparator, examining just one square inch at a time, spending every night photographing the sky.

On 18 February 1930 Tombaugh was examining a pair of plates when he saw a speck of light jump as he compared the images. It was a tiny difference, scarcely one-third of a centimetre (one-seventh of an inch). Tombaugh's observation was confirmed by astronomers and on 13 March 1930 – the anniversary of Lowell's death – the observatory announced the discovery of Planet X. There were many suggestions for a name for the new planet. That of an 11-year-old English girl named Venetia Burney was chosen because "Pluto", the Roman god of the underworld, seemed appropriate – and because its first two letters were also Percival Lowell's initials.

Beyond Pluto

Are there any planets beyond Pluto? That was the first question asked as soon as the discovery of Pluto was announced, and it is a discussion that refuses to go away.

At the time that Clyde Tombaugh discovered Pluto, most astronomers thought that even if there were more distant bodies out there, it would be all but impossible to find them. It had been difficult enough to discover Pluto and few if any astronomers were willing to dedicate the time and effort to do what Tombaugh had done simply in order to find a speck of light no larger than a pinpoint. In the decade following his discovery, Tombaugh spent another 7,000 hours at the blink comparator. He found asteroids,

Neptune's Orbit

Kuiper Belt

Oort Cloud

LEFT: Beyond the orbit of Neptune lies the Kuiper Belt, a region of icy, Pluto-like worlds. Further out is the vast Oort Cloud, a reservoir of countless billions of small, icy bodies that may be the source of the comets that occasionally swing into our inner solar system.

OPPOSITE: The view from the frontiers of our solar system. Out beyond Pluto lie the frigid worlds of the Kuiper Belt, where the Sun is little more than an especially bright star in the black sky.

more than 1,800 variable stars, nearly 30,000 galaxies and even a comet. But no Planet 10. He finally concluded that "no unknown planet beyond Saturn exists that is brighter than magnitude 16.5. ..." But many astronomers began to wonder.

We know today that the planet Pluto is very small, even in comparison to our own diminutive planet. It is only 2,370km (1,473 miles) in diameter, about two-thirds the size of Earth's Moon. The small scale of the planet in itself raised a new question for consideration. Pluto is obviously too small to have caused the disturbances that had been seen in the orbits of Uranus and Neptune. However, if Pluto hadn't caused them, what did? Perhaps the real Planet X has actually not yet been discovered.

The possible presence of a large planet out beyond Pluto involves a mystery concerning comets. Our solar system is surrounded by a swarm of comets called the Oort Cloud, named after Jan Oort, the Dutch astronomer who discovered it in the 1950s. The cloud averages 50,000AU (astronomical units) from Earth, which is more than a thousand times that of Pluto's distance from the Sun. Most of the well-known comets that enter the inner solar system from time to time, such as Halley's Comet, Shoemaker-Levy 9 or Hyakutake, originate in the Oort Cloud. But what starts them on their inward spiral toward the Sun? Is there something out there disturbing the comets' routine orbits? Could it be an unknown planet?

"The presence of at least one large planet out beyond Pluto is a near certainty."

Ron Miller

"If they be inhabited, what a scope for misery and folly; if they be uninhabited, what a waste of space."

Attributed to Thomas Carlyle (1795–1881)

OPPOSITE: The New Horizons spacecraft flew past Pluto in 2015. Now speeding through the Kuiper Belt, its Earth-bound crew hopes to steer it around a Kuiper Belt object, providing scientists with their first close-up view of one of these strange worlds.

The discovery that there are other astronomical bodies similar in size to the planet Pluto orbiting beyond it also gives credence to the possibility that a much larger body may exist in the distant frontiers of our solar system. The region just beyond Pluto, called the Kuiper Belt, is between about 30 and 55 times farther away from the Sun than Earth. It is a scientifically fascinating zone, populated with hundreds of thousands of icy bodies larger than 96.5km (60 miles) across. Some of these are sizable objects indeed, occasionally nearly as large as Pluto itself. Sedna, for example, discovered in 2004, is about three-quarters the size of Pluto. Eris is nearly the same size as Pluto but is more than twice as far away from the Sun.

In 2016 researchers at the California Institute of Technology (Caltech) announced that they had found evidence suggesting the presence of a possible Neptune-sized planet orbiting on the farthest fringes of our solar system, about 20 times farther from the Sun on average than Neptune, or nearly 600 times farther away than Earth. This object might have a mass about ten times that of Earth and take between 10,000 and 20,000 Earth years to make one full orbit around the Sun. Note that no one has yet imaged this planet directly. Instead its existence was inferred by its gravitational effects on other objects in the Kuiper Belt in just the same way that Adams and Le Verrier had assumed the presence of Neptune by its effect on Uranus.

IS EARTH UNIQUE?

In 1710 German mathematician Gottfried Leibniz said that our planet must be the best one anyone could imagine. The French satirist Voltaire lampooned this idea in his novel *Candide* (1759). A few years earlier he dealt with the same theme in his classic story "Micromégas" (1752), in which he described a giant alien visitor from the planet Sirius who learns that humans believe that the entire universe was uniquely created for the sole benefit of mankind. The idea that all the billions of planets, stars and galaxies of the universe circle around tiny Earth and its even tinier inhabitants causes the huge alien to burst out in a fit of laughter. Yet, 300 years later many religions make the same arguments and even

some scientists have expressed similar doubts – that if life exists elsewhere, it is an extremely rare phenomenon. The Rare Earth Hypothesis, championed by palaeontologist Peter Ward, astrophysicist Howard Smith and astronomer Donald Brownlee, is based on the premise that extreme conditions are likely to be the norm on other planets and that the hospitable conditions we find on Earth might be unique. However, in the 15 years since the book *Rare Earth* (2000) was published, discoveries in our own solar system, as well as the discovery that Earth-like planets may be much more common than originally thought, have actually made the theory seem far less tenable.

ABOVE: Voltaire (1694–1778) was a French writer, historian, philosopher and satirist who was especially fond of attacking religious dogma, often in relation to early thoughts about Earth's place in the universe.

ALIENS ON EARTH

Centuries ago, the new worlds being discovered on our own planet were as strange and mysterious as Mars or Venus seemed to be. Efforts to imagine what weird types of creatures might inhabit these places mirror modern attempts by science fiction authors and flying saucer enthusiasts to visualize the inhabitants of other planets.

BELOW: Having little knowledge about what sorts of beings inhabited Earth's then-unknown continents, European artists gave free reign to their imaginations. These medieval illustrations depict creatures described by Pliny the Elder in his *Natural History* (77–79 AD).

  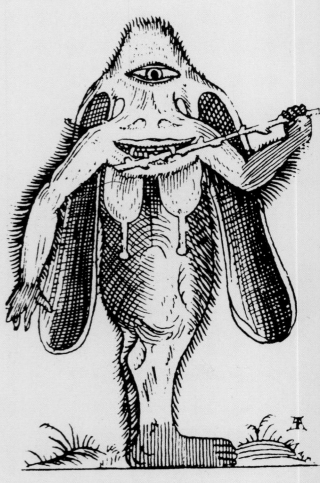

"Life, may, indeed, take on many different visages throughout the universe."

Michael Carroll (2001)

BELOW LEFT & RIGHT: Medieval artists and writers filled books, pamphlets and broadsheets with incredible creatures. In 1478, Conrad von Megenberg depicted some of the strange people he believed lived in distant, unexplored regions of the world (below left). In 1475, an unknown artist filled an entire page with the weird creatures he thought might inhabit those lands (below right).

EARLY IDEAS ABOUT EXTRATERRESTRIAL LIFE

OPPOSITE: In Louis-Guillaume de la Folie's 1775 book *Philosophe sans Pretention*, an inhabitant of Mercury pays a visit to Earth in an electrically propelled flying machine.

ABOVE: Johannes Kepler (1571–1630) not only discovered the basic laws that govern the motion of the planets, but he was also among the first people to write a science fiction story that included aliens.

The Christian Church attempted to censor Galileo's findings in the first decades of the 17th century, but it was a time of expanding knowledge, so it did not take very long for the information to get out. When Galileo's observation became widely known, people started wondering if these other worlds were like our own. Did life exist on them? Did people live there? Even the Church finally decided that such speculation was not blasphemous. As the truth of the plurality of worlds became accepted, it was assumed that God would never knowingly create a world for no reason.

It was decided that if other worlds in space *did* exist, their only purpose could be to provide a home to humanlike creatures. As Thomas Burnet asked in his book *The Sacred Theory of the Earth* (1681):

> God himself formed the Earth . . . he formed it to be inhabited. This is true, both of the Earth and of every habitable World whatsoever. For to what purpose is it made habitable, if not to be inhabited? We do not build houses that they should stand empty, but look out for Tenants as fast as we can.

It did not take long before several books were published speculating about what sort of life might exist on the planets. Some authors assumed that any sentient life existing on the other planets would necessarily have to be humanlike. Other authors took a looser definition of what constituted "human", with the idea that what was most important was the quality and nature of the *mind*, not the form of the shell that bore it.

Journey into Speculation

The great German astronomer Johannes Kepler wrote what might be the first science fiction novel, *Somnium*, which was published in 1634 (a few years after his death). As a serious scientist, he described the Moon and the sort of creatures that might live on it as accurately as the knowledge of the time allowed. The Moon was an incredibly alien world, he told his readers. Nights were 15 Earth days long "and dreadful with uninterrupted shadow". The cold at night was more intense than anything experienced on Earth, while the heat of day was terrific. Animals that lived on the Moon adapted to these harsh conditions. Some went into hibernation, while others evolved hard shells and other protection.

As the 17th century progressed, the concept that the planets were not only inhabitable but

inhabited was taken for granted. In 1656 the Jesuit priest and writer Athanasius Kircher, sent the hero of *The Ecstatic Journey* touring the heavens with an angel as his guide. In the course of these journeys through the celestial world, the Moon was actually found to be quite habitable, with a varied terrain that included mountains, oceans, lakes, islands and rivers. In *Paradise Lost* (1667), John Milton has the angel Raphael and Adam discussing the possibility of life on other worlds, including the Moon. But the angel cautions Adam, saying that it is dangerous to think about such matters since God does not intend human beings to comprehend everything about his creation: "Dream not of other worlds, what creatures there live, in what state, condition or degree."

A French mathematician named Bernard de Fontenelle was not afraid to dream about such things and wondered what sort of creatures might exist on the planets in his book *Entretiens sur la Pluralité des Mondes* (*Conversations on the Plurality of Worlds*; 1689). In fact, he not only asked the question, but he also attempted to answer it. And he did so in a way that was completely unique. De Fontenelle explained that although the planets were worlds very much like our own, conditions there would probably be vastly different. For example, Mercury would be incredibly hot because it is so close to the Sun. If life existed on the planets it would necessarily have to reflect an adaptation to these very specific conditions.

ABOVE: Bernard de Fontenelle's *Entretiens sur la Pluralité des Mondes* (1689) was one of the first books to speculate seriously about the possibility of life on other worlds and what forms that life might take. De Fontenelle was careful to take into account the physical conditions on other planets and how that might influence the life forms that could be existing there.

"This Earth? reciprocal, if Land be there,
Fields and inhabitants? Her spots thou seest
As clouds, and clouds may rain, and rain produce
Fruits in her softened soil, for some to eat
Allotted there; and other Suns, perhaps,
With their attendant Moons…"

Milton (1667)

"Then there will also be people, who do not shrink from the dreary vastness of space."

Johannes Kepler (1609)

RIGHT: These two creatures were encountered by the hero of *Nicolai Klimii Iter Subterraneum* (*Niels Klim's Underground Travels*), an early science fiction novel written by Ludvig Holberg in 1741. In the story, Earth is discovered to be hollow, with a small planet called "Nazar" orbiting within it. Holberg's hero, Niels Klim, finds that it is inhabited by an entire menagerie of bizarre beings, including tree people and denizens of "the land of music".

Figure d'un Botuan.

Une Habitant du païs de Musique.

The problem de Fontenelle faced was the simple fact that scientists didn't know enough. All he had to work with were the approximate sizes of the planets and their approximate distances from the Sun. Other than being able to make rough estimates of the planets' surface temperatures based on their relative distances from the Sun, he knew nothing at all about the nature of the conditions on them. He had no way of knowing what a planet's atmosphere might be like – or even for sure if it had one. Still, de Fontenelle didn't let a little impediment like that inhibit his imagination and he proceeded to describe the creatures that lived on the other planets in great detail. The inhabitants of Mercury, he declared, were exuberant, excitable and quick-tempered. They "resemble the Moors of Granada, a small, black people, burned by the Sun, full of wit and fire, always in love, writing verse, fond of music, arranging festivals, dances and tournaments every day."

The people of Venus, by contrast, were incorrigible flirts, those of Jupiter were great philosophers and the denizens of Saturn, because of the frigid climate of their planet, preferred to sit in one place for their entire lives. However, de Fontenelle decided the Moon was probably uninhabited due to its thin atmosphere.

Between 1764 and 1772, Italian artist Filippo Morghen published *Raccolta delle cose più notabili veduta dal cavaliere Wilde Scull, e dal sigr: de la Hire nel lor famoso viaggio dalla terra alla Luna* (usually known in English by the shorter title, *Land of the Moon*). The book was a series of etchings depicting the fantastic creatures and landscapes of Earth's satellite. Here, Morghen shows us "A Savage Mounted on a Winged Serpent Battling With a Wild Beast Resembling a Porcupine".

OPPOSITE: An illustration from the original publication of the "Moon Hoax" story depicts many of the supposed lunar inhabitants.

"To consider the Earth as the only populated world in infinite space is as absurd as to assert that in an entire field sown with millet, only one grain will grow."

Metrodorus of Chios (4th century BC)

In *A Voyage to the World of Cartesius* (1694), Gabriel Daniel describes the inhabitants of the Moon as being entirely spiritual, without physical bodies at all, who can travel from place to place by the force of their will alone.

Ralph Morris' *A Narrative of the Life and Astonishing Adventures of John Daniel* (1751) tells of the invention of a machine that carries a shipwrecked sailor on a trip to the Moon. Once there he discovers copper-skinned humanoids who live in caves and worship the Sun. In *Le Philosophie sans Prètention* (*The Unassuming Philosopher*; 1775) by Louis-Guillaume de la Folie we learn that the hero, Ormisais, has flown to Earth from the planet Mercury. He informs the Earthlings he meets that a Mercurian scientist named Scintilla has invented an electrical flying machine capable of travelling between worlds. The hero of *A Voyage to the Moon* (1793), Aratus, travels to the Moon by balloon ("facilitated by air currents"), where he finds a race of English-speaking snakes who walk upright on legs.

Books like these – both fanciful and realistic – helped convince their readers that other worlds did exist and that it was possible there might be life on them. People even began to wonder if the stars themselves might be other suns. After all, they asked, if the universe were infinite in size – as was supposed at the time – it should have an infinite number of stars. Wasn't it reasonable to suppose that at least some of them might be suns like our own? And if they were suns like ours, might they not also have planets circling them?

AN INHABITED MOON?

When the Great Moon Hoax was perpetrated by the *New York Sun* reporter Richard Adams Locke, it created enormous public excitement about the possibility of life beyond Earth on other worlds, our own Moon in particular. Published in 1835, it was probably inspired by astronomer Franz von Paula Guithuisen's announcement in 1822 that he had observed a walled city on the Moon near the crater Schröter (*see* p.192). Locke told his readers of startling discoveries being made by the British astronomer Sir John Herschel at his observatory at the Cape of Good Hope in South Africa. Herschel was the son of William, who had passed away in 1822 but was still world famous for his discovery of the planet Uranus. Dropping his name lent Locke's articles an air of believability. Needless to say, Herschel himself, who was in South Africa at the time, had no idea that his name was being bandied about so freely.

Locke explained that Herschel had invented a kind of super-telescope built on entirely new principles. Through the use of a "hydro-oxygen microscope", he was able to magnify and project the telescopic image onto a large screen. When he did this, Herschel discovered humanoid, fur-covered, bat-winged beings living on the Moon. Readers thrilled to Locke's descriptions of these bizarre creatures, as well as giant crystals, bison and weird plants — to say nothing of beavers that walked on two legs and lived in primitive huts.

Anticipating the literary technique of verisimilitude (of seeming true to life) that was picked up by Edgar Allan Poe (who was inspired to write his own Moon travel story) and through Poe, Jules Verne and every science fiction author since, Locke used plausible detail and scientific-sounding jargon to convince his American readers of the reality of his reports. To allay any remaining doubts, Locke assured readers that "several Episcopal, Wesleyan, and other ministers . . . were permitted, under stipulation of temporary secrecy, to visit the observatory, and become eyewitnesses of the wonders which they were requested to attest."

The hoax became enormously popular in Europe, where it was translated and reprinted in French, German and Italian, inspiring numerous artists to create prints and lithographs showing the imaginary discoveries in great detail. All of this interest was perhaps encouraged by the reported presence of nude Moon-maidens.

The *New York Sun* finally confessed the whole thing was a hoax. When Sir John eventually arrived back home and learned how his name and reputation had been so freely used, he took it with great good grace and humour, remarking that his only regret was that he would never be able to live up to the fame.

The First Science Fiction Aliens

Camille Flammarion, a French astronomer and science popularizer, was a firm believer in extraterrestrial life. In *Real and Imaginary Worlds* (1865), he declared that the planets were inhabited, but that the beings living there were repositories for human souls after the death of their bodies on Earth. In *Lumen* (1897), Flammarion discussed at length how the conditions existing on a planet influence the development of life there, even that of our own world: "The human organism is the product of the planet. It is not by a Divine fantasy, by a miracle, or by a direct creation that terrestrial man is constituted such as he is. His form, his figure, his weight, his sense, his whole organization, are derived from the state or condition of your planet, the atmosphere that you breathe, the food that

nourishes you, the gravity of the surface of the earth, the density of terrestrial matter, &c."

A mixture of mysticism and theology with speculation about extraterrestrial life persisted throughout the 19th century in such books as the Reverend W S Lach-Szyrma's *Aleriel* (1883), John Jacob Astor's *A Journey in Other Worlds* (1894), James Cowan's *Daybreak* (1896), George Griffith's *Honeymoon in Space* (1900) and William Arkwright's *Utinam* (1912). In these and other novels, aliens were revealed to be anything from transmogrified souls to angels to the person-ification of ancient gods. With the possible exception of Flammarion, most of these authors were less interested in the reality of alien life as they were in using the idea as a platform for expounding their particular views about philosophy or religion.

ABOVE: In 1850, an artist created a series of vivid colour pictures depicting a visit to the Moon. Here we see a lunar landscape that includes a pair of centipede-like creatures, one of whom might have been the progenitor of Lewis Carroll's hookah-smoking caterpillar in the famous novel *Alice's Adventures in Wonderland* (1865).

ABOVE: A cartoon published on the cover of *Puck* magazine in 1909 depicted Martians as elfin humanoids.

RIGHT: The Martians discovered in Fenton Ash's 1909 novel *A Trip to Mars* were typical of the era in being depicted as idealized – even angelic – human beings.

However, as the 19th century progressed and knowledge about the planets increased, both scientists and science fiction authors had to revise their ideas about life on other worlds. Whatever the creatures that lived on the other planets might look like, they probably didn't resemble human beings.

But while science fiction's conception of extraterrestrial life was slowly evolving, science showed little interest in pursuing the possibility of life on other worlds. The more astronomers learned about the true conditions on the other planets, the less likely it seemed that life might exist there. Even Mars, it was thought, could boast nothing more advanced than scant patches of mosses and lichens, struggling to exist in the thin, frigid air. Mercury was either too hot or too cold to support life, depending

ABOVE: The birthplace of our solar system: an artist's depiction of a cloud of interstellar dust and gas collapsing on a massive scale under its own gravity, growing ever denser and hotter as it does so.

"We conclude, then, that our Earth
is a highly distinguished planet..."

Scientific American (1873)

RIGHT: The proto-solar system: a disk of dust and gas with
a hot, dense core in which the Sun is soon to be born.

on which side of the planet you were on. Jupiter, Saturn, Uranus and Neptune were bitterly cold with atmospheres of poisonous gases. Venus was the only mystery. What possibly lay beneath its opaque blanket of clouds? A steaming prehistoric jungle; a vast, planet-wide ocean; or a barren desert? Astronomers could only shrug their shoulders and guess.

The Birth of a Solar System

While we know there are planets circling our own star and that life may have evolved on a few of them – and even more especially on their moons – the questions remain. Is our solar system unique or are there worlds circling other stars? Are planets common or rare? To begin to answer those questions, we first have to understand how our own solar system came into being.

The currently accepted explanation for the evolution of our solar system – and Earth – goes something like this: 100 million lifetimes ago, what was to become the Sun and planets was nothing but an enormous cloud of dust and gas, now known as the presolar nebula. It was the cocoon in which our Sun was born. Perhaps encouraged by the shockwave from a nearby supernova, this cloud began to collapse and condense under the weight of its own gravity. As big as it was to start with, the cloud shrank rapidly to a millionth of its original size.

As it shrank, the centre of the cloud became denser and its gravity increased. This, in turn, made it collapse even further and faster. The tremendous, ever-increasing pressure at the centre of the cloud caused its core to heat up, at first with just a dull, red glow. After only a few

"Our search for earths of distant suns has deep roots ... from the biological imperative to simple curiosity"

Michael Carroll (2016)

thousand years, the temperature and pressure became great enough to trigger a nuclear reaction – and the protostar became a star. The increased amount of heat this produced created an outward pressure that resisted the continuing collapse of the dust and gas.

Within the cloud, tiny particles of dust collided and stuck together, forming little clumps of material called planetesimals. By chance, one of these was slightly larger than the rest, which gave it an advantage. Its growth became even more rapid. As it grew in size, it attracted even more particles in a process called *accretion*. It gradually became the size of a rock, then a boulder and then a planetoid many miles across. By the time it became about 322km (200 miles) wide, it was starting to take on a spherical shape. The whole process of growth to this point may

have taken only about 100,000 years. By now the original supply of dust and gas was getting used up. The greedy planetoid became cannibalistic, devouring one smaller body after another and becoming ever larger. The infant Earth began to grow even more rapidly, achieving its present size in as little as 40 million years.

Planets of Other Stars

If the accretion theory is correct, then solar systems may be commonplace throughout the universe. All evidence seems to show that this may be true. The Hubble Space Telescope has observed dozens of infant solar systems in almost every phase of development. But infant solar systems are one thing. Are there, in fact, planets around other stars? Until modern times, astronomers assumed that this would always be

OPPOSITE: The infant Earth. The original gas and dust from the formation of the solar system coalesced into particles, which in turn became larger particles as they collided. Drawn together by their mutual gravity, these grew into ever greater planetesimals and eventually became planet-sized bodies.

LEFT: As a planet passes in front of the star it orbits (a phenomenon called a "transit"), the level of light from the star will decrease slightly. From these variations in a star's brightness, the existence of the planet can be inferred.

"There are innumerable worlds of different sizes . . .
some are flourishing, others declining."

Democritus (4th century BC)

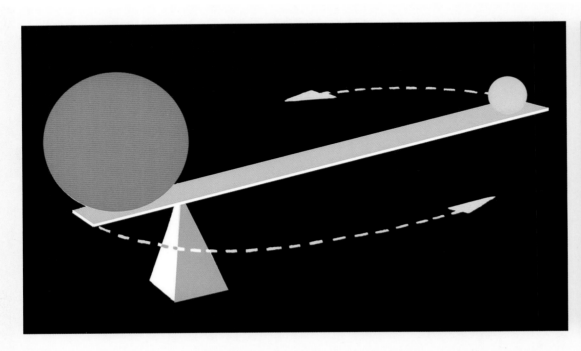

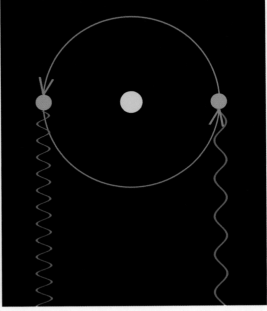

impossible to determine. Planets are both tiny and dark – they give off no light of their own – two factors that would make extrasolar planets extremely difficult to detect from Earth. But astronomers have other ways of discovering things beyond than just by direct observation with telescopes.

The existence of a planet might be inferred from the effect of its gravity on its star during the orbit. As the planet swings around the star, it pulls first one way and then another, causing the star to wobble ever so slightly on its axis. The star is rather like a hammer thrower in the Olympics as he swings his hammer around his body in a circle.

In addition to detecting the wiggle an invisible companion can impart to a star, the existence of a planet might be discovered if its orbit causes it to pass in front of its star. When this happens, the star's light is dimmed slightly. A third

technique for detecting objects in the sky depends on the fact that planets give off more infrared radiation than stars do. When the astronomers use a telescope sensitive to infrared radiation but insensitive to visible light, they are able to detect the presence of a planet so near to its star that it would normally have been lost in the glare.

Finally, astronomers can detect planets around other stars by looking for the Doppler effect or shift. This effect is the change in pitch you hear in the sound of a train whistle or police siren as the vehicle speeds past you. As it approaches, its sound seems higher pitched than when it's going away. This is because the wavelength of the sound is compressed as the vehicle comes toward you and stretched as it moves away.

In the same way, the wavelength of light is changed by the motion of the object emitting the light. When a star is approaching you, its

ABOVE LEFT: As a planet orbits a star, it causes the star to wobble slightly – like a pair of spinning, unequal weights might. When astronomers detect such a wobble, they can infer the presence of one or more planets.

ABOVE RIGHT: Scientists can also detect the presence of a planet by means of the Doppler effect. This term describes the phenomena that when a body moves toward or away from Earth, the wavelength of its light is compressed toward the blue or lengthened toward the red. A planet tugging a star toward it and away from Earth will cause the star's light to become redder or bluer, from which astronomers can assume the presence of a planet.

ABOVE: A brown dwarf is neither a star nor a planet, but rather something in between. It is so large that it radiates enough energy to glow a deep red, but it is not large enough to trigger the fusion that would transform it into a true star.

wavelength is shortened slightly, making the light seem a little bluer (because blue light has a shorter wavelength). When the star is moving away, the wavelength of the light is increased, making the light seem a little redder (because red light has a longer wavelength). Using an instrument called a spectrograph, astronomers can easily detect these changes in the wavelengths of light and, in turn, measure the

speed that distant objects are moving toward or away from Earth. The measurement of Doppler shift changes has been the primary method used to date in the search for extrasolar planets. So, using this method, astronomers have been able to extend their search for life beyond our own solar system. But before that can really reap results, we need to know what we are looking for: just what is life?

VICTORIAN ALIENS

Whenever extraterrestrials made an appearance
in 19th-century literature, it was almost
invariably assumed that they would be humanoid
— and not only humanoid but often superhuman
to the point of being virtually angelic. It was the
rare author who dared suggest that aliens might
be genuinely *different*.

LEFT: H G Wells created the first genuinely non-human
aliens in *The War of the Worlds* (1898), here depicted by
artist Alvim Corrêa (1876–1910) as lusting after a human
female in true alien fashion.

BELOW: For most of the 19th century — and well into the
20th — aliens were represented as idealized humans, such
as these inhabitants of Ganymede from *A Honeymoon in
Space* by George Griffith (1900).

TOP LEFT: In *Voyage dans la Lune avant 1900* (*A Trip to the Moon Before 1900*), a children's book published in 1890, artist-author A de Ville d'Avray has his heroes encounter many strange inhabitants on our satellite.

BOTTOM LEFT: Visitors to Mars in Arnould Galopin's *Le Docteur Oméga* (1906) had to run the gauntlet of imaginative monsters, including bat-winged humanoids.

BOTTOM RIGHT: The explorers visiting Mars in *1500 Miles an Hour* (1895) by Charles Dixon discover this "hideous amphibious monster". Like many authors at the end of the 19th century, Dixon tried to shape aliens to reflect the conditions then known to exist on the planets. In this case, he ascribed to the theory that Mars was covered by extensive marshes.

02

"We are limited in our definitions of life and of intelligence. We have only one example in the form of our own planet."

Michael Carroll (2016)

THE SCIENCE OF EXTRA-TERRESTRIAL LIFE

THE BIOGRAPHY AND BIOLOGY OF A PLANET

It is difficult to define "life" since we Earthlings have only one example to go by: the things that inhabit our own planet. As wildly diverse as terrestrial life might seem – from octopuses to daisies – it is all in fact closely related. Every living thing on our planet evolved from the same simple, one-celled organisms, which is one of the reasons we all share to a greater or lesser degree the same DNA – the chemical code that allows life to pass characteristics from one generation to the next.

So before we could identify something on another planet as being alive, we would first have to recognize it as a living creature. There are complex organisms on our planet that at first glance don't even resemble animals at all. For example, a species of sea squirt, *Pyura chilensis*, looks exactly like an ordinary rock. Many of us might agree with NASA (the American National Aeronautics and Space Administration) – which has an abiding interest in the subject, given its investment in the search for life on other worlds. In a piece from *Astrobiology Magazine* on its website NASA suggests that:

Living things tend to be complex and highly organized. They have the ability to take in energy from the environment and transform it for growth and reproduction. Organisms tend toward homeostasis: an equilibrium of parameters that define their internal environment. Living creatures respond, and their stimulation fosters a reaction-like motion, recoil, and in advanced forms, learning. Life is reproductive, as some kind of copying is needed for evolution to take hold through a population's mutation and natural selection. To grow and develop, living creatures need foremost to be consumers, since growth includes changing biomass, creating new individuals, and the shedding of waste. To qualify as a living thing, a creature must meet some variation of all these criteria.

However, many scientists – and the NASA organization itself – realize that there may not be a hard and fast definition of what constitutes "life", and that it may be counterproductive to even look for one. There may not even be a sharp dividing line between "life" and "non-life". By way of comparison, imagine how Earth's atmosphere slowly thins as it gets higher and higher above the surface of the planet, until it eventually merges seamlessly into the vacuum of outer

space. But where does the transition from "atmosphere" to "no atmosphere" actually take place? It's really just a gradation, like the grey scale between white and black, and the point where "life" becomes "non-life" might be just as impossible to pinpoint.

It is conceivable that viable life forms – even intelligent life forms – may exist that do not possess all of the attributes NASA lists. The intricate matrices in a crystal, for example, may emulate the neurons in your brain and perhaps do so even more efficiently. Dr Carol Cleland, a member of NASA's Astrobiology Institute, explains: "I don't think that defining 'life' is a very useful activity for scientists to pursue since it is not going to tell us what we really want to know, which is 'what is life.' A scientific theory of life (which is not the same as a definition of life)

would be able to answer these questions in a satisfying way."

The Birth of Life

Just as we need to understand how our solar system – and Earth – came about in order to understand how other planets may have formed, to understand how life might emerge on another world in space we have first to understand how it came to exist on Earth. It is, after all, the only example that we have.

Between 4.2 and 3.5 billion years ago, the last of the debris left over from the formation of our solar system had been swept up by Earth, the Moon and the other planets. What had been a chaotic period of almost endless impacts was now settling down to a more peaceful routine. As asteroid collisions became less frequent,

ABOVE: Our planet would have been better named "Ocean". Water covers seven-tenths of its surface, drives its weather and is home to more than 90 per cent of the life forms that inhabit our world.

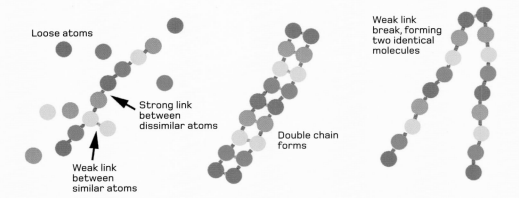

Loose atoms

Strong link between dissimilar atoms

Weak link between similar atoms

Double chain forms

Weak link break, forming two identical molecules

ABOVE RIGHT: Long, chain-like molecules with strong and weak links between the atoms can replicate themselves. Relatively simple versions of these molecule chains soon evolved into more complex replicating molecules. These became the DNA that drives evolution and is still the foundation for life on our planet.

Earth was able to form a stable crust. At this time Earth was a lifeless wilderness of black rock surrounded by broad oceans filled with silt eroded from the fledgling continents by constant rain and the enormous tides raised by a newly formed moon.

Within that murky ocean an extraordinary event was taking place. Earth's primordial seas were a complex soup of minerals and elements, as well as vast amounts of dissolved carbon dioxide (CO_2). These oceans, rich with nutrients and gases, infused with the energy provided by sunlight, lightning and tides, were ultimately to become the cradle of life.

Carbon – one of the elements that forms carbon dioxide – has the ability to produce large, complex molecules. It can do this in combination with a great many other elements, such as hydrogen,

nitrogen and oxygen. Another, and perhaps even more important property, of many carbon-based molecules is their ability to split into two nearly identical halves. Each half then absorbs the necessary elements to recreate the original molecule almost exactly. In other words, they can reproduce themselves. This relatively simple, but incredibly important, ability of a molecule to reproduce is the foundation of all the life that exists on Earth today.

No one knows when the first living cells evolved from these early, reproducing molecules nor even exactly how that happened, but we know that by at least 3.5 billion years ago enough true cells existed to begin leaving traces in the fossil record. These have been found in some of the oldest rocks on our planet, dating from 2 to 3.2 billion years ago: the dawn of life itself.

"The mystery of the beginning of all things is insoluble by us; and I for one must be content to remain an agnostic . . ."

Charles Darwin (1887)

ARE *YOU* AN ALIEN?

Beyond our own solar system, scientists know that complex organic molecules are commonplace throughout the universe. They can be found in a wide range of cosmic locations from vast nebulae to the icy tails of comets. This suggests that the stars themselves may be seeding infant solar systems with organic compounds and, consequently, there is the possibility of life on those planets. Closer to home, scientists have found abundant evidence of the basic building blocks of life – such as carbonaceous materials, amino acids and organic compounds – in certain types of meteors and in comets. We also know that Earth was regularly bombarded by comets and asteroids during the period of its formation. One theory is that much of the water on Earth today came from the impact of millions of icy comets. Is it possible that these impacts could have also seeded Earth with amino acids, proteins and perhaps even viruses, from which all present-day life on this planet evolved? If so, then you yourself may be a distant descendant of life brought to Earth from the stars.

OPPOSITE: Earth's nutrient-rich primeval oceans – with energy provided by sunlight, tides, lightning and other sources – became the cradle for life on our planet. In this artist's depiction we see primitive algae sheltered in a coastal tidal pool. Even today, algae continue to produce most of the oxygen found in Earth's atmosphere.

The First Living Things

The first life forms to inhabit Earth were bacteria and their relatives, the blue-green algae. They are among the simplest of all organisms, their cells lacking even a nucleus. This structure, however, left the DNA in the cells unprotected and vulnerable to damage, and damaged DNA cannot reproduce accurately.

Between 2 and 3 billion years ago, cells began to evolve a nucleus: a kind of protective container for the fragile DNA. Two types of these improved cells eventually evolved. One type resembled the original algae and was the ancestor of today's plants. The other was a more complex version that evolved into animals including, eventually, human beings.

The original atmosphere of Earth had largely been made up of carbon dioxide and water vapour. But with the ever-increasing abundance of blue-green algae, oxygen began to dominate the atmosphere. The algae was using sunlight to break down atmospheric carbon dioxide in order to obtain the carbon it needed to build organic molecules. The excess oxygen was released into the air and eventually replaced most of the carbon dioxide in the atmosphere.

The fact that there is any oxygen in our atmosphere at all is due entirely to the efforts of Earth's plant life, which constantly replaces the oxygen that is lost through combustion and other chemical reactions. Earth's original atmosphere was as little as 1 per cent oxygen and more than 90 per cent CO_2 – much like present-day Mars and Venus. Today, oxygen accounts for about 21 per cent of Earth's atmosphere and CO_2 only 0.039 per cent. It took many millions of years to create enough oxygen to raise it to that level.

An animal life requires much more energy than plant life. An animal has to search for its own food – while avoiding becoming food – and doing this requires a constant source of energy. This energy comes from combining fuel with oxygen and without a source of oxygen, complex animal life – such as ourselves – cannot exist. So, ironically, it was a substance discarded by ancient plants as a waste product that we can thank for making the evolution of animal life – and eventually human beings – possible.

An Explosion of Life

Over the next 150 million years there was an explosive proliferation of life forms of all kinds, both in the plant and animal kingdoms. Primitive one-celled animals quickly evolved into complex multi-celled organisms: animals. By 550 million years ago, animal life flourished so abundantly that fossils from this period are found on every continent. Between 450 and 400 million years ago, the first fish swam in Earth's seas and life finally began moving to dry land. At first this consisted of little more than simple mosses and lichens, but eventually the continents were covered with vast forests of giant ferns.

These great forests blanketed the world of 320 million years ago. Giant insects shared the

ABOVE: Life soon established itself on dry land in the form of primitive ferns, lichens and mosses. These proliferated, creating vast forests that covered the continents.

OPPOSITE: While plant life was spreading over the surface of the world, simple algae were replacing the early carbon dioxide-dominated atmosphere with one that consisted chiefly of oxygen: a necessary step toward the appearance of animal life.

"In eons past, life on Earth has started, been erased from the surface, and repopulated from biota beneath its surface."

K B Kofoed (2016)

LEFT: Reptiles and dinosaurs dominated Earth for tens of millions of years. These creatures were among the most successful forms of life ever to have evolved on our planet, although they remained under threat from external forces, such as volcanic eruptions. It is only by the action of a random event that they do not continue to exist and evolve to this day.

land with the ancestors of today's frogs and salamanders, which in turn had evolved from lobe-finned lungfish. These early amphibians were the ancestors of the reptiles that eventually came to dominate Earth's early history for tens of millions of years.

Then, about 250 million years ago, a mysterious and global catastrophe wiped nearly 90 per cent of all species from the face of the planet. The trilobites and most of the amphibians vanished. And of the fifty genera of mammal-like reptiles, only one remained. It was from this survivor that all of the modern mammals – including human beings – evolved.

No one is certain what caused the "Great Extinction" or "Great Dying". Theories range from sudden climate change to radiation from a

nearby supernova. The disaster was a boon for the survivors of this catastrophe, as they now had far less competition. Reptiles proliferated wildly, and from 250 million years ago to 65 million years ago they dominated the world.

The 185 million-year-long success story of the dinosaurs eventually came to an end. This terminus was probably the result of the impact of an asteroid 9.6km (6 miles) wide that smashed into a region of Mexico just off the coast of present-day Yucatán. An explosion equal to 100 million one-megaton hydrogen bombs instantly turned both the asteroid and the earth beneath into an incandescent cloud of vaporized rock and steam, creating a crater nearly 201km (125 miles) wide as debris erupted high into the stratosphere. This debris spread into a world-encompassing

ABOVE: The "Great Dying" that occurred 250 million years ago wiped out 90 per cent of the life that existed on Earth. Only by sheer chance did one tiny creature survive to become the distant ancestor of every human being alive today.

cloud that blocked the light from the Sun, possibly for decades.

Meanwhile, millions of tons of red-hot debris began to fall back onto the earth, igniting forest fires all over the world. These raged for months, destroying life and habitats and pouring vast clouds of smoke and ash into the already sunless sky. As the sky grew darker, Earth consequently grew colder. Without sunlight, plants died. And without plants to eat, animals died. Seventy-five per cent of all the species on Earth did not survive this endless winter.

It took millions of years, but the hardy survivors of the asteroid impact slowly evolved into new kinds of animals. Without the competition of the giant lizards, tiny, warm-blooded, furry, placental animals — what we now call mammals — began to

flourish. Mouse-sized, omnivorous tree-dwellers similar to today's tarsiers and lemurs, they were an inauspicious harbinger of the creature they were to become 10 million years later. They were the first steps on the road to humankind.

The distant forebears of human beings — small, chimpanzee-like creatures that walked upright — seem to have appeared 3 or 4 million years ago. Although their brains were only about half the size of a modern human's, they were using tools made of bone and stone. They gradually and inexorably spread from Africa into Europe and Asia, evolving all the time, becoming more complex, more intelligent, more capable. Our present species, *Homo sapiens*, evolved only about 200,000 years ago, a mere tick of the cosmic clock.

The Lesson

The history of life on Earth is a 4.5 billion-year-long history of accidents, millions upon millions of them – and if any one of them had taken another path, life on our planet might have been very different than we know it today. You might not even be here to appreciate that.

The fundamental processes of the evolution of living organisms might be the same on every world that even remotely resembles our own. The right chemicals might be there, the right temperatures and the right amount of water, but once the first organic molecules come into being, there are an endless number of paths they can take. There are nearly 2,600,000 different five-card possible poker hands in a deck of 52 cards. There are 20,000 genes in the human genome – imagine how many different ways *those* could be arranged.

Yet, in spite of all the physical and biological differences between the living things on Earth, every living organism on this planet shares a significant amount of DNA with every other organism. You, as a human being, share 98 per cent of your DNA with chimpanzees, 85 per cent with zebrafish, 36 per cent with fruit flies and even 7 per cent with bacteria. You even share about a not-inconsiderable 25 per cent of your DNA with rice and grapes!

What this shows is that all life on Earth has a common, shared ancestry. Life beginning from scratch on another world would be dealt a fresh hand from the gene deck. And in the first hand dealt it would have to match – gene for gene – almost everything that occurred on Earth, including billions of random mutations and other events, in order for life there to be like life on our

"It is not reasonable to think that all this immense vastness should lie waste, desert, and uninhabited, and have nothing in it that could praise the Creator thereof, save only this one small spot of Earth."

Ralph Cudworth (1678)

own planet, let alone evolve intelligent creatures that resemble human beings. It also might be worth pointing out that human beings are not a finished product. It is easy to think of ourselves as the end result of millions of years of evolution, but we are in reality just a work in progress. Just 200,000 years ago we were an entirely different species. And 60 million years ago we were a small, shrew-like proto-primate. And a billion years before that we were one-celled organisms swimming in a primordial sea. What will we be 200,000 years in the future? A million? Sixty million? Whatever those creatures might be, they will be as far removed from us as we are from our most distant ancestors.

One has only to marvel at the incredible diversity of life on just this planet alone to appreciate how wildly different the paths that life on another world might have followed, and over what timelines. Such an appreciation highlights how naïve it is to expect that the evolution of alien life might have followed those same paths to the same conclusions – and did so following exactly the same timeline, era by era. But this is what would have had to have happened if we believe the stories we hear from people who have met aliens that resemble human beings in any significant way.

"ALIEN" LIFE ON EARTH: THE EXTREMOPHILES

Many of the worlds in our own solar system were once thought to be too hostile to support any kind of life even if all the necessary materials were at hand. Mars, for example, is intensely cold and its surface is bathed in lethal ultraviolet radiation from the Sun. Io (Jupiter's fifth moon) is intensely volcanic, with vast lakes of molten sulphur and lethal radiation, this time from Jupiter. Europa (Jupiter's sixth moon) and Enceladus (Saturn's sixth moon) may have oceans of liquid water, but they lie miles beneath a thick icy crust through which not a photon of light can penetrate However, a discovery was made here on Earth that suggests the possibility of life existing on these and other worlds. This was the discovery of life existing in extremely hostile environments such as hot springs, acidic ponds and salt marshes – environments even more hostile than anything that might be expected on Mars or many other worlds in our solar system. These organisms, rightly called extremophiles, live in areas on Earth that are so salty, acid, alkaline,

ABOVE: These shrimp thrive in the hostile environment near a hydrothermal vent, from which mineral-laden water is spewing at temperatures of up to 400°C (750°F).

poisonous, hot or cold that it is hard to imagine familiar living things being able to survive. Many scientists now believe that if life can evolve and even flourish in such unlikely conditions on Earth, it might well be possible for life to exist on worlds once thought too hostile to support any kind of organism.

WHEN ALIENS INVADED EARTH

Five hundred million years ago there was an explosion of life on Earth that resulted in some of the most bizarre creatures to ever inhabit this world. On both land and sea, and in the air, these creatures would not be out of place in any science fiction movie. Creatures known as the "Burgess Shale fauna" (after the location in Canada where they were discovered) were an evolutionary dead end, related to no known species living today.

BELOW AND OPPOSITE: The evidence of the Burgess Shale shows the almost infinite forms life can take. It is only by chance that our world today is not as alien as that in any science fiction story. Among the bizarre and even nightmarish creatures were (19) Pseudoarctolepis, (7) Anomalocaris, (8) weird Opabinia with its five eyes and nozzle nose, (29) Dinomischus, (21) Marella, (22) Branchiocaris, (24) Hyolithes and the aptly named Hallucigenia (5, also below). Other creatures of the Burgess Shale included (1) stromatolites, (2, 3) sponges, (4) brachiopods, (6) lobopods, (9, 10, 11, 12, 13) trilobites, including the odd (14) Naraoia, (15, 16, 17, 21, 22) arthropods, (20) the crayfish-like Canadaspis, (23) Ottoia, which preyed upon mussels like (24) Hyolithes, (25) segmented worms, (27) wormlike creatures, (26) primitive sea lilies, and utterly unclassifiable animals such as (28) Wiwaxia, (29) Dinomischus and (30) Amiskwia.

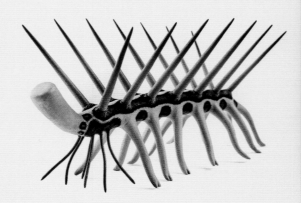

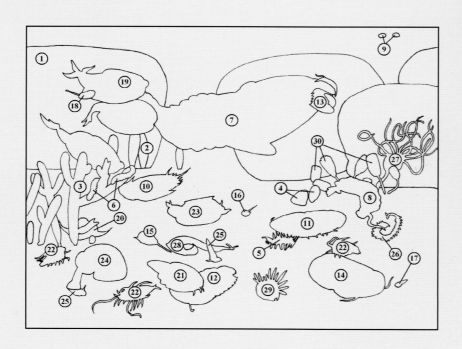

04

IS THERE LIFE ELSEWHERE IN THE UNIVERSE?

OPPOSITE: Artist Richard Bizley's painting of jellyfish-like creatures floating high among the clouds of an alien world is evocative not only of the vast possibilities of life on other planets, but also the inspiration it has long given to artists and authors.

BELOW: The rich orange, yellow and red colours that dominate the thick, stormy clouds of Jupiter are due to complex organic compounds in its atmosphere. Here, the artist has allowed us to plunge deep within the planet's turbulent clouds.

After decades of denying any possibility of life elsewhere in our solar system – with the scant exception of Mars – scientists in recent years have come to realize that many of these worlds probably aren't as hostile to life as once thought. Mercury and Venus are still considered too hot to support life, while the outer planets – Uranus, Neptune and Pluto – are probably too cold. But even these seemingly unwelcoming worlds may not be lifeless, as scientists discover just how tenacious life can be. Take, for example, the recent discovery of amino acids – essential building blocks of life – on comets. In light of this, scientists have not only been taking another look at Mars, but Jupiter and Saturn as well.

The brilliant orange, red and yellow colours of Jupiter's clouds are due entirely to the presence of organic molecules, ranging from carbon dioxide to ammonia. While these compounds are themselves not living, they are carbon-based molecules from which more complex molecules can evolve. The visible surface of Jupiter's swirling clouds is much too cold to support life and deep beneath them conditions are even more hostile, with crushing pressures and temperatures as high as 19,427°C (35,000°F). But somewhere in between the core and the upper clouds, temperatures and pressures would be conducive to the evolution of life. Are there living balloons and zeppelins cruising like whales through the towering clouds of Jupiter? Do airborne jellyfish float through Jupiter's atmosphere, like those once suggested by Carl Sagan in a paper he co-published with fellow astronomer E E Salpeter in 1977?

Deep beneath the 19km (12-mile) thick icy crust of Jupiter's moon Europa lies a vast ocean of warm water that may be as deep as 96.5km (60 miles). This moon has an abundance of salts and organic materials, and plenty of energy from the heat created due to the flexing of Europa's mass by Jupiter's powerful tides. What life forms

might exist in those utterly lightless depths? The bizarre creatures that live in the abysses of Earth's oceans might provide a clue. So convinced are astronomers that life may teem in the Europan sea that serious work has gone into planning a robotic mission to Europa to drill through the ice in search of life.

Other moons in our solar system may be home to life forms as alien as anything we can imagine. Saturn's giant moon, Titan, has a dense atmosphere rich in organic compounds. Its

surface is literally buried beneath a thick slurry of organic materials while rivers of liquid methane flow into hydrocarbon seas. Even Saturn's tiny Enceladus has a hidden ocean of warm liquid water rich in salts and organic molecules.

Curiosity About Life on Other Worlds

Most early writers were aware that the special conditions on other worlds would produce creatures unlike those on Earth. In 1698 Dutch scientist Christiaan Huygens, knowing that Mars

ABOVE: The surface of Europa is laced with thousands of cracks and fissures in its icy crust. The brown stains are caused by organic compounds that have welled up from the deep, warm ocean that lies beneath.

"Many of the solar system's moons may be home to lifeforms beyond our understanding or imagination."

Ron Miller

RIGHT: The possibility of life existing beneath Europa's ice has engendered much science fiction, from the film *Europa Report* (2013) to author Allen Steele's novelette *Angel of Europa*, which inspired this artwork.

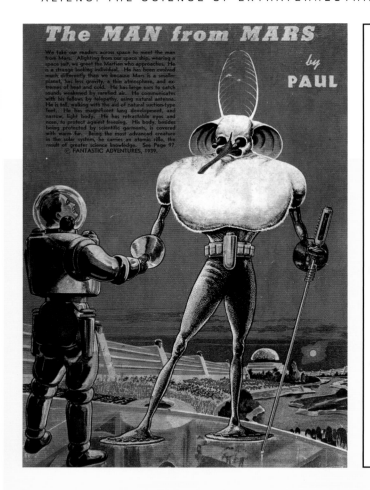

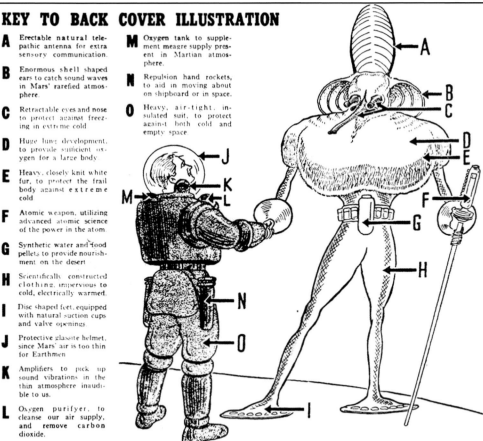

is much farther from the Sun than Earth and therefore probably a colder world, suggested that Martians would be covered in fur and feathers. The Swedish mystic and philosopher Emanuel Swedenborg was a firm believer in the habitability of other worlds, saying: "That there are many earths, and men upon them." In his 1758 book, *Earths in the Universe*, he explained that Martians would be gentle beings who dressed themselves in the bark from trees. He described the inhabitants of the Moon as dwarfs about the size of a seven-year-old boy, though stouter in stature and with thunderous voices. Mercurians resembled Earthlings and wore tight clothing. They were also hungry for knowledge, able to read minds and counted Aristotle as one of their own. The inhabitants of Venus were divided into two groups, one peaceful and gentle, the other fierce thieves. The inhabitants of Jupiter were upright, soft-spoken, happy and family-oriented people, who were obsessed with washing their

faces and had a tendency to walk on their hands. Finally, Saturnians were restrained, humble people with low self-esteem. They had little interest in food and clothing. Rather than burying their dead, they would cover them with leaves.

The anonymous author of the *Fantastical Excursions into the Planets* (1839) believed that the size, mass, gravity and climate of the other planets suggested a likewise wide variety of possible life forms. William Herschel believed that even the Sun itself might be inhabited.

In spite of the work of these and other writers, most people before the turn of the 20th century assumed that Martians and other aliens would be more or less human in appearance. It took H G Wells' horrifying novel *The War of the Worlds* (1898) to introduce the idea of monstrous, inhuman, hostile aliens.

But why pick on Mars? Of all the planets in our solar system, why has Mars traditionally been the first suspect when people talk about life beyond

ABOVE, LEFT & RIGHT: In 1939 classic science fiction artist Frank R Paul depicted a Martian according to the best information about the planet available at the time. For example, he allowed for the thin atmosphere and intense cold of Mars. To this end, he gave his Martian enormous lungs and a covering of thick fur. He deemed that huge ears would be necessary to pick up sound in the tenuous atmosphere and telescoping eyes that could be retracted when it got too cold.

"The amazing blue network on Mars hints that one planet besides our own is actually inhabited now."

Percival Lowell (1894)

RIGHT: This map by Italian astronomer Giovanni Schiaparelli (1835–1910), depicting what he termed "canali", gave rise to Percival Lowell's assumption that the lines crisscrossing the planet were the result of intelligent creatures living on Mars.

Earth? Even the phrase "little green man from Mars" has become a cliché signifying any alien creature. This is probably because among all the planets, Mars seems to be most Earth-like. It has an atmosphere, polar caps, clouds and – at least to early viewers – signs of some sort of vegetation in the seasonal changing of colour. Even the Martian landscape would not seem all that unfamiliar to a visitor from Earth, especially one who has ever travelled to Iceland, the Chilean Atacama or the American Southwest.

The Amateur

It was the theories about Mars proposed by Percival Lowell that affected how generations of writers, scientists and ordinary people perceived the red planet. In 1877 the Italian astronomer Giovanni Schiaparelli announced that he'd observed the surface of Mars crisscrossed by a network of thin lines. This was unusual enough, but it was the word Schiaparelli used to describe

these lines that startled the world. He called them *canali*. In Italian this means simply "channels" or "grooves", but it is so similar to the English word "canal" that everyone assumed that was what Schiaparelli was talking about. The difference – and it was a crucial one – is that *canali* could describe a natural feature, while *canal* refers only to an artificial structure.

The debates began immediately. On the one side were the astronomers who couldn't see anything resembling "canals" on Mars and denied that they even existed. On the other side were those who *did* see them. This latter group quickly divided into two camps: those that thought the "canals" were natural features and those that insisted they were artificial.

Percival Lowell fell squarely and enthusiastically into the second group. A decade before he began his search for Planet X, he had dedicated his observatory to the observation of Mars and the mapping of the red planet. The maps Lowell

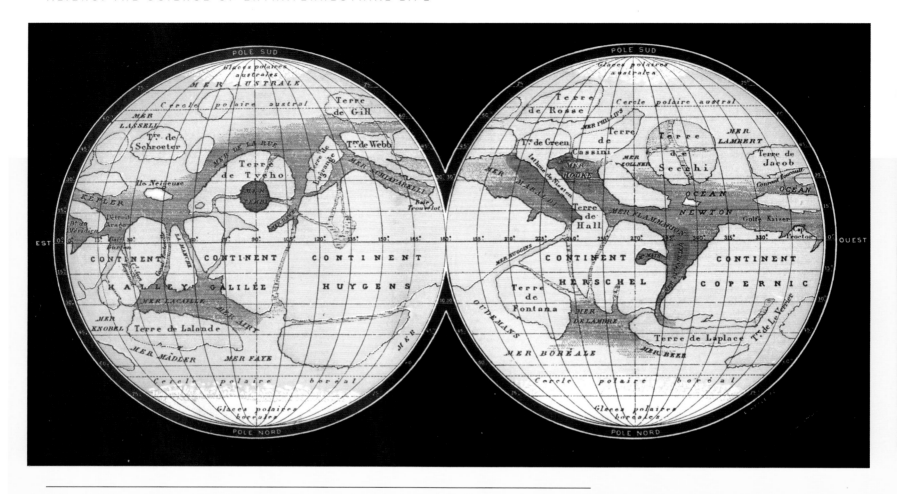

ABOVE: Other astronomers had different ideas about how to interpret the markings on Mars. British astronomer Richard Proctor (1837–88) thought the dark areas were rivers, lakes and seas.

"Since stars appear to be suns, and suns, according to the common opinion, are bodies that serve to enlighten, warm, and sustain a system of planets, we may have an idea of the numberless globes that serve for the habitation of living creatures."

William Herschel (1795)

created were unlike anything seen before. They were covered with a dense network of straight, thin, dark lines. They looked as perfectly artificial as a modern map of airline routes. To promote his theories about Mars, Lowell lectured widely and published two popular books, *Mars* (1895) and *Mars as the Abode of Life* (1908), in which he vividly expressed his ideas about the true nature of the "canals". (It might be worth mentioning that Lowell saw canals on *everything*, including

Mercury and Venus.) "We saw," he wrote, "how badly off for water Mars, to all appearance, is; so badly off that inhabitants of that other world would have to irrigate to live. . . . How to procure water enough to support life would be the great communal problem of the day."

Lowell's Mars was an ancient desert world, much older than Earth, that had lost most of the water it might once have had. (Lowell turned out to be right in one respect. Modern astronomers

THE HURTLING MOONS OF BARSOOM

Although what little science existed in Edgar Rice Burroughs' series of novels set on Mars was out of date even at the time he was writing, his books proved to be immensely influential. The American astronomer Carl Sagan expressed this debt when he said: "The Mars novels of Edgar Rice Burroughs . . . aroused generations of eight-year-olds, myself among them, to consider the exploration of the planets as a real possibility" (*Cosmos*; 1980). In *Listen to the Echoes* (2010), Ray Bradbury agreed, saying:

"I've talked to more biochemists and more astronomers and technologists in various fields, who, when they were ten years old, fell in love with John Carter and Tarzan and decided to become something romantic. Burroughs put us on the moon." To several generations of young readers, it hardly mattered that the Mars described by Burroughs not only never existed but probably *couldn't* have existed. What mattered was that the red planet was described vividly, with a breathless, matter-of-fact realism

ABOVE: A modern artist's interpretation of the dying red planet, as described by Edgar Rice Burroughs in his Barsoom series.

that overwhelmed the imagination. It is hardly any wonder that scores of scientists and astronomers were inspired to discover for themselves what that mysterious world was really like. Nor is it any wonder that today a large crater on Mars is named "Burroughs".

believe that Mars had oceans, lakes and rivers in its distant past. Today much of the water that has not been lost to space is frozen deep beneath the surface.) The surviving Martians were digging a vast system of canals to carry water from the polar caps to the central regions of the planet. Lowell never claimed that it was the actual canals themselves he saw – they would have to have been many miles wide to be visible from Earth. Instead, what he was seeing was the

broad belts of vegetation that flourished alongside the canals, similar to how the lush Nile valley appears from space to today's astronauts.

Astronomers looked on these announcements with a combination of scepticism, disdain and annoyance, but the public was enthralled. Many authors and their readers were inspired by Lowell's romantic description of a dying world, its few remaining inhabitants struggling against a planet-wide drought by heroically digging vast

> "Mars, therefore, is not only uninhabited by intelligent beings such as Mr. Lowell postulates, but is absolutely uninhabitable."
>
> Alfred Russel Wallace (1907)

canals that crisscrossed their world, carrying the last drops of water from the shrinking poles. Everything from novels, short stories and magazine articles to plays, games, fairground rides and popular songs – even a march and two step called "A Signal from Mars" was published – reflected the appeal this tragic race of beings had on the over-sentimental Victorian mind.

Lowell was never specific about what he thought Martians might look like. But in keeping with expectations of the time, science fiction authors who were inspired by Lowell assumed that Martians would be more or less human, an idea taken to its extreme limits by Edgar Rice Burroughs in his justly famous Barsoom series (Barsoom was the Martian native name for the red planet in the books), which began with *A Princess of Mars* in 1914. Set in a distant past

when Mars was still largely a going concern, Burroughs' planet was populated by numerous different races… all of which were entirely human even if they did lay eggs, like platypuses.

H G Wells' Martians

H G Wells was one of the few authors who took seriously the question of what form Martian life might take and how that would be driven by its alien environment. His masterpiece in this line was, of course, the classic science fiction novel *The War of the Worlds* (1898). In fact, there were two current fads that Wells drew upon in the writing of his book. The first was the novels about future wars that were immensely popular at the time. The second was, of course, the public's fascination about Mars, stirred to a fever pitch by Lowell's books. Wells combined these two

PANSPERMIA

Panspermia comes from a Greek word meaning "seeds everywhere" and refers to the theory that the basic constituents for life exist throughout the universe. These "seeds" – perhaps in the form of complex molecules or even DNA – are so small that the pressure of starlight alone can propel them through space. Although this would be a slow process, over countless eons they could have spread across our galaxy and perhaps even between galaxies. There are scientists who have suggested that life here on Earth originated in such seeds

drifting down to our planet. Although the basic idea goes back at least as far as the Greek philosopher Anaxagoras, it was first popularized by the Swedish chemist Jöns Jacob Berzelius (1779–1848), the British physicist Lord Kelvin (William Thomson) (1824–1907) and the German physicist Hermann von Helmholtz (1821–94). The theory lay more or less dormant for decades, with only a handful of scientists advocating it, but today, with the discovery that complex organic compounds exist throughout the universe, the concept has had a revival.

The idea of directed panspermia suggests that these seeds were sent into the universe *deliberately*. This opens up the possibility that we are all alien invaders, distant descendants of chemical seeds that fell to Earth billions of years in the past. The idea can work both ways, of course. It has been suggested that instead of sending humans to the stars in enormous spacecraft on journeys that may take generations, human DNA could be scattered into space to ride from world to world on currents of starlight.

"... the Martian stems and stalks will all be slenderer and finer and the texture of the plant laxer ... it seems reasonable to expect bigger plants there than any that grow upon the Earth."

H G Wells (1908)

OPPOSITE: Illustrator Charles Dudouyt created this interpretation of H G Wells' monstrous Martians for an early French edition of *The War of the Worlds*.

ABOVE, LEFT AND RIGHT: Depictions of H G Wells' octopoid Martians by two different artists. (Above left) British artist Warwick Goble illustrated the first edition of *The War of the Worlds* in 1897. (Above right) In 1906 Brazilian artist Alvim Corrêa tried his hand at Wells' Martians. His creatures were certainly the more malevolent and horrifying of the two depictions.

strands, at the same time working in a serious commentary on British imperialism. Just as Wells took the themes of future war and life on Mars to their logical extremes, he also turned the accepted ideals for life on other planets upside down. Instead of glorified humans, Wells depicted aliens as being truly *alien*. His cold-blooded, octopoid Martians were the product of a world with a lower gravity than Earth — making it difficult for them to move while on our planet and forcing them to resort to machines for transport. They were like nothing ever before encountered by the reading public. Moreover, unlike the benevolent, humanoid aliens found in earlier stories, these creatures were patently evil — "intellects vast and cool and unsympathetic" — with no other goal than the systematic extinction of humanity, excepting the handful of humans who they planned to retain as slaves. But there was still one thing lacking in Wells' aliens. They *looked* unearthly, but we never really get to learn anything about their minds. An intelligent alien might look very different from a human being, but would it also *think* differently?

"We shall find no flies nor sparrows nor dogs nor cats on Mars."

H G Wells (1908)

Taking Lowell's canal theory at face value, Wells wrote a nonfiction article for *Cosmopolitan* magazine in 1908 called "The Things That Live on Mars" in which he tried to describe life on the red planet as realistically as possible. In fact, this essay may be among the earliest serious thoughts in modern times regarding what extraterrestrial life might actually be like:

Here then is one indication for a picture of a Martian animal: it must be built with more lung space than the corresponding terrestrial form. And the same reason that will make the vegetation laxer and flimsier will make the forms of the Martian animal kingdom laxer and flimsier and either larger or else slenderer than earthly types.

And now we are in a better position to consider those ruling inhabitants who made the gigantic canal-system of Mars, those creatures of human or superhuman intelligence, who, unless Mr Lowell is no more

than a fantastic visionary, have taken Mars in hand to rule and order and cultivate systematically and completely, as I believe some day man will take this earth. Clearly these ruling beings will have been evolved out of some species or other of those mammal-like animals, just as man has been evolved from among the land animals of this globe. . . .

There are certain features in which they are likely to resemble us. The quasi-mammalian origin we have supposed for them implies a quasi-human appearance. They will probably have heads and eyes and backboned bodies, and since they must have big brains, because of their high intelligence, and since almost all creatures with big brains tend to have them forward in their heads near their eyes, these Martians will probably have big shapely skulls. But they will in all likelihood be larger in size than humanity, two and two-thirds times the mass of a man, perhaps. That does not mean,

ABOVE LEFT: Writers of the 19th century drew inspiration from Percival Lowell's highly romanticized descriptions of a dying race bravely and desperately eking out an existence on a declining world, their cities clustered around oases in endless tracts of flat, arid desert.

ABOVE RIGHT: William R Leigh's illustrations for H G Wells' 1908 article about life on Mars included this charming family group. The low gravity of the planet accounts for the delicate bodies of the Martians and the spindly architecture.

ABOVE: The first photographs of the actual surface of Mars, taken by the Viking landers in 1976, told a different story. Under salmon-pink skies were revealed planet-wide deserts with no sign of water, let alone canals.

however, that they will be two and two-thirds times as tall, but, allowing for the laxer texture of things on Mars, it may be that they will be half as tall again when standing up. And as likely as not they will be covered with feathers or fur. Even if astronomers were somewhat less than convinced about the likelihood of canal-building Martians, the idea that life of *some* kind existed on Mars certainly seemed plausible.

The Search Continues

The debate among scientists was not *whether* there was life on Mars but rather what kind of life it might be. It was the ongoing conviction that even if there was no life on Mars at the time, it must have existed sometime in the past that

drove the launch of the twin Viking probes in 1976. When the landers touched down in the Martian plains Chryse Planitia and Utopia Planitia, many scientists were so certain that they would detect signs of life that what actually occurred came as a complete surprise.

The twin landers were specifically created to search for life, with five experiments designed to detect signs of biological activity. The results raised more questions – and started more debates – than they answered. Two of the experiments were negative. The cameras and soil analyzer were made to look for the remnants of life, that is, signs that life might have existed at some time in the planet's past. The images relayed back to Earth revealed nothing that

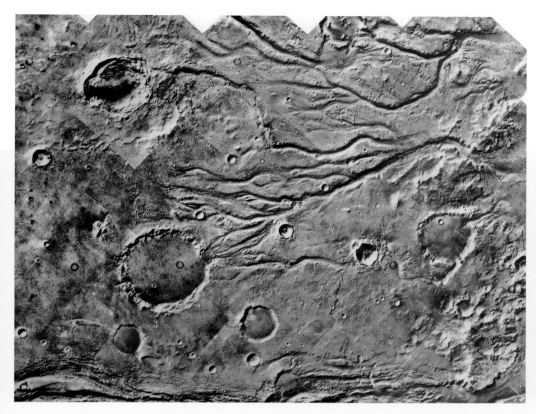

OPPOSITE: Mars is a geophysically complex world, hosting the solar system's largest mountains, greatest canyon and, possibly, vast reservoirs of water beneath its rust-coloured deserts.

ABOVE LEFT: Viking orbiter images shocked scientists when they revealed clear evidence that, while it may appear to be arid today, water actually flowed abundantly on Mars' surface in the past.

ABOVE RIGHT: Images taken by later orbiters showed that Mars might not be as arid as once thought. Many features, such as these channels, appear not only to have been created by running water, but moreover seem to have been made recently.

suggested that life had ever existed on Mars and the soil analyzer discovered no organic molecules.

However, the other three experiments were designed to look for ongoing biological activity, such as the products of photosynthesis. To do this, samples of soil were placed into special chambers where nutrients were added. These were then monitored. All three tests returned positive results. Did this mean there are living organisms on Mars? Many scientists thought so.

The excitement was short-lived. The possibility was raised that ordinary chemical processes and not life had produced the readings. So while Viking didn't find any conclusive evidence for life on Mars, neither did it prove that there was no life on the planet, or that life hadn't existed there in the past. After all, it was reasonably pointed out, Mars has more dry land area than Earth and the Viking landers were able to explore only an incredibly small part of this. It was pointed out that Viking-like tests had failed to find signs of life in Antarctic soil samples, where micro-organisms were known to exist. Mars has a very thin atmosphere that allows most of the Sun's

deadly ultraviolet radiation to reach the surface. Perhaps life on Mars has merely retreated underground.

But all was not lost. The photos that the Viking orbiters took of the Martian surface revealed some totally unexpected – and visually spectacular – geological features. Winding channels, braided streambeds and teardrop-shaped islands were incontrovertible evidence that water had once flowed on the surface of Mars. But how much water? For how long? And, most importantly, where did it go?

The Search for Water

Scientists know that Mars has all the necessary ingredients for life. Or at least it once did, millions of years ago. The planet was warmer and had liquid water. It had all the same elements, minerals and chemical compounds as Earth did when life first formed here. Also like the early Earth, Mars had energy from volcanoes and the Sun. Billions of years ago, Mars also had vast lakes and maybe even oceans. These seas were probably a mix of minerals and elements washed

"Today, rock 84001 speaks to us across all those billions of years and millions of miles. It speaks of the possibility of life."

President Bill Clinton (1996)

from the land by rain and meteor impacts. The water also probably had vast amounts of dissolved carbon dioxide.

Scientists know that on Earth, the billions of different complex molecules that formed in its ancient oceans were the building blocks from which more complex life forms developed. For this reason, scientists wonder if complex, self-reproducing molecules may have formed in the carbon-rich chemical waters of the early Martian seas.

One exciting proof of water on ancient Mars appeared in 1984, when scientists exploring Antarctica found an extremely rare type of meteorite that comes from Mars. Labelled ALH 84001, the 2.26kg (5lb) chunk of rock had been blasted nearly 4 billion years ago from a region on Mars now called Eos Chasma, when an asteroid impacted the planet. After a long journey through space, it collided with Earth about 13,000 years ago.

Martian meteorites all contain traces of gases and other elements that resemble conditions on the red planet at the time the meteorites were formed there. When ALH 84001 was launched into the vastness of space, scientists believe that there were seas and lakes of liquid water on the surface of Mars. So, scientists reasoned that the meteorite might actually contain evidence of ancient life on Mars.

In examining the meteorite, scientists found not only traces of Mars' ancient atmosphere, but globules of orange carbonate that resembled the deposits of limestone found in caves on Earth. This material forms about 1 per cent of the meteorite. This was a significant find on many levels. Since limestone can form only in the presence of liquid water it meant that at some time in the past the original rock had sat in water. Additionally, the globules were covered with microscopic worm-like shapes that resembled fossil bacteria. However, one objection to this is that the shapes are so small that they would have to be considered "nanobacteria".

It was then found that these globules contained traces of organic compounds as well as iron sulphide and magnetite. While magnetite can be produced by ordinary geological processes, the kind found in ALH 84001 is a chemically pure form that is known to be produced only by magnetotactic bacteria. Many scientists at the time doubted that these were signs of ancient Martian life and they gave examples of ways in which these materials could have been created by non-living processes. But whether or not ALH 84001 contains Martian

ABOVE: Found in the Antarctic, a meteorite labelled ALH 84001 contains what some scientists believe is evidence that primitive life once existed on the surface of Mars. This evidence takes the form of microscopic objects that closely resemble types of bacteria found on Earth.

ᴬᴮᴼᵛᴱ: NASA's Mars Science Laboratory Curiosity rover, seen here in an artist's rendering, has provided solid proof that water once existed in abundance on the surface of Mars. Curiosity has been exploring the region of Gale Crater since 2012.

fossils, it was among the first proofs that water once flowed on the red planet – and illustrated vividly how deeply ingrained in us is the desire to find life there.

The Mars Science Laboratory

The Mars Science Laboratory – also known as the Curiosity rover – landed on Mars in 2012. Among its goals is to determine if life ever existed on Mars. It wasn't specifically designed to find life still living now on Mars (other than what it might spot with its cameras).

The Curiosity rover was designed to discover if there were ever conditions on Mars that might have supported life in the past. Even if that life had been nothing more than simple microbes, this would be an important scientific discovery. It would prove that life is indeed capable of forming on worlds other than our own.

Although signs of ancient life have still eluded the rover, it has succeeded in confirming that billions of years ago Mars was a very wet planet. It had long been known that water once flowed on Mars – there are channels and teardrop-shaped islands to demonstrate that – but it was thought that surface water might have existed for only a short time. Too short, perhaps, to have aided in the evolution of life. Now scientists know that Mars had long-lived streams and lakes sometime around 3.8 to 3.3 billion years ago. There is even some evidence that Mars had extensive oceans. The lake that once filled Gale Crater, which Curiosity is exploring, may have been up to half a mile deep and lasted as long as half a billion years. The challenge facing scientists now is to understand not just how this clement Mars was possible in the first place but also to understand what happened to it.

"The present inhabitation of Mars by a race superior to ours is very probable."

Camille Flammarion (1892)

OPPOSITE: In its distant past, Mars may have been a watery world, as imagined in this image, with a climate conducive to the evolution of life. However, a weak gravitational field and erratic climate shifts contributed to the eventual loss of much of its original atmosphere and, consequently, its water.

RIGHT: In 2008 NASA's Phoenix Mars lander confirmed the existence of underground water ice. The white patches in this image are ice revealed by the lander's robotic scoop. The mission also found perchlorate, a chemical that is a food for some microbes on Earth.

Where Is the Water Now?

Mars may have had seas, lakes and rivers millions of years ago, but where did the water go? And does water still flow anywhere on Mars today?

It was generally assumed that whatever water Mars once had existed either entirely of ice, as small amounts of water vapour in the atmosphere, or was buried hundreds or even thousands of metres underground. While they might have existed in the distant past, there are no lakes or rivers on Mars today. The atmosphere is largely to blame for this – or at least the lack of it. There simply isn't enough pressure at the surface to keep water from rapidly evaporating. And what didn't evaporate froze. It's possible that much of Mars' ancient water was lost to space. It was the warmer temperatures and denser atmosphere Mars possessed 3.8 billion years ago that allowed liquid water to flow on the surface of the planet.

As for liquid water existing on Mars today, a few scientists held out hope, but it seemed like little more than wishful thinking. Then in late 2015, NASA's Mars Reconnaissance Orbiter photographed dark, narrow, hundred-yard-long streaks running down the slopes of hills and craters in many different locations on Mars. They appeared to darken and flow down steep slopes during the Martian summer, and then fade when it became colder.

The streaks were assumed to be formed by seasonal flows of water occurring today. The spacecraft's spectrometer detected hydrated salts on the slopes, corroborating the hypothesis that the streaks are being formed by briny liquid

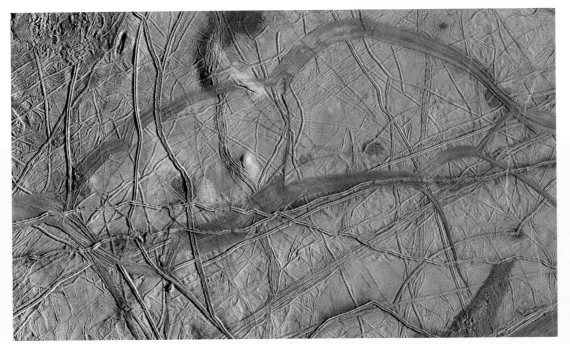

LEFT: In 2015 NASA announced plans to launch a probe to Europa. The orbiter will use an ice-penetrating radar to search for subsurface lakes as well as a magnetometer, which will allow scientists to determine the depth and salinity of the water. This image taken by the Galileo orbiter shows the fractured surface of Europa. Its thick, icy crust is constantly shifting – like ice floes in the Arctic here on the Earth – and as this crust cracks, water laden with organic materials wells up to the surface. This organic material is what has left the orange stains on the surface.

water. The brininess of the water explains how it has been able to flow even during the subfreezing temperatures that are more or less the norm on Mars. The concentration of salts dissolved in the water lower its freezing point, just as salting a winter road on Earth causes ice and snow to melt more rapidly.

Until this discovery, it was assumed that all of the water on Mars was either frozen or buried deep underground. Scientists now know – after nearly two centuries of searching – that liquid water exists on Mars today and regularly flows over its surface. And where there is water, there may be life.

Is There Life Elsewhere in Our Solar System?

Where scientists once restricted their search for alien life to the most Earth-like of all the planets, Mars, many are now beginning to believe that there are other worlds in our solar system that might not only harbour life but may be even more likely to do so than the red planet.

The prime candidates are Europa (one of Jupiter's four largest moons), Enceladus (a tiny moon of Saturn), Titan (Saturn's planet-sized moon) and even giant Jupiter itself. Both Europa and Enceladus are known to have large subterranean reservoirs of water. Even though Europa is a small world – it is slightly smaller than Earth's Moon – its ocean may contain two to three times as much water as all of Earth's oceans combined and could be up to 100km (62 miles) deep.

In addition to water, we also know that Europa possesses two of the other three ingredients necessary for the evolution of life: a source of energy and raw materials. The latter consists of complex, organic, carbon-based molecules that are the building blocks of life. Scientists know these chemicals exist on Europa because they can literally see them. Thousands of linear cracks lace its otherwise smooth surface, making Europa resemble, if nothing else, Percival Lowell's old maps of Mars. These cracks are caused by Europa's miles-thick icy crust shifting above the sea of liquid water below. It is exactly the same process that creates ice floes in the Arctic and Antarctic ice sheets on our planet. As the ice cracks, water from deep beneath the surface erupts through the fissures, carrying with it the organic compounds it contains. The result is that Europa's cracks are lined with reddish-brown stains – a colour that is one of the hallmarks of

56   POPULAR SCIENCE MONTHLY   MAY, 1930

Do Beavers Rule on Mars?

No trace of human intelligence has been found on the red planet, and it is thought that evolution, through lack of the stress that helped on earth, may have halted with some animal adapted to a land and water life.

By THOMAS ELWAY

MARS is so like the earth that men might live there. It has air, water, vegetation, a twenty-four-hour succession of day and night, and daily temperatures no hotter and nights not much colder than are known on earth. But because Mars has no mountain ranges and probably never had an Ice Age, it is considered highly improbable that it is inhabited by manlike creatures or by any that possess what men call intelligence. The evolution of life on Mars must have been different from that on earth.

One of the best signs of intelligence on Mars, Dr. Clyde Fisher, of the American Museum of Natural History, New York City, said recently, would be some indication of artificial light on the planet. Undoubtedly, lighted cities on Mars could be seen through the telescopes now in use.

However, there is one condition that prevents satisfactory and conclusive observation. When Mars is closest to the earth, both planets are on the same side of the sun. Then only the sunlit side of Mars is seen. To see any part of the night side of Mars, observation must be made when it is part way around in its orbit toward the far side of the sun, so that a slice of both the dark and the lighted sides can be seen.

When even a part of the night side is visible, Mars is relatively far away and difficult to see clearly. The Martians, if there are any, would not have equal difficulty in observing the dark side of the earth, for when the two planets are nearest to each other, the earth is showing Mars its dark side.

THESE consequences of the orbits in which the two planets move might make it difficult for the dim glow of lighted Martian villages, were any such in existence, to be detected from the earth. Cities as bright as New York or Paris, on the other hand, undoubtedly would be visible. With the new 200-inch telescope which, it is planned, will be erected in California, it surely would be possible, Dr. Fisher predicted, to distinguish such brightly lighted cities, if any such Martian centers of civilization exist. If such artificial lights are never seen, he added, it might go a long way toward proving that Mars does not possess intelligent life. Other students of the subject, however, say it is possible that Martian

civilization may correspond to that of an earlier, pre-artificial light era on earth. In any case, astronomers agree that there is a practical certainty that Mars possesses kinds of life below human intelligence. Any deduction about the life forms on Mars or other planets, in the opinion of leading astronomers, must start, if it is to be at all reasonable, with the idea of the distinguished Swedish scientist, Dr. Svante Arrhenius, of one kind of life-germ pervading the entire solar system. There is no reasonable way even to guess the form of this life-germ. It may, perhaps, have drifted, as tiny living spores, from planet to planet, whirled through space by the pressure of light.

Whatever its form, the life-germ, biologists assume, probably developed on Mars, much as it did on earth, in oceans which have evaporated in the course of ages. Early conditions on the two planets are supposed to have been very similar.

The theory that Martian life evolved along lines similar to those followed by evolution of life on earth is supported by at least one definite fact. Careful spectroscopic studies at Mt. Wilson Observatory, near Pasadena, Calif., and elsewhere have disclosed that gaseous oxygen exists in the Martian atmosphere. The presence of oxygen gas is highly significant, since the only known way in which any planet can obtain a supply of this gas is through the life activities of plants.

Following the lead of the great expert in Martian astronomy, the late Professor Percival Lowell, astronomers long have recognized on Mars dark-colored spots and markings which are believed to be plains or valleys covered with vegetation. The oxygen which spectroscopes show in the Martian air is taken as another proof that this vegetation exists.

Since the activity of plants is the only known process of cosmic chemistry by which free oxygen can be produced on the surface of a cooled planet, the presence of oxygen in the rarefied air of Mars indicates that vegetation there must have produced oxygen out of water and sunlight as it has done on earth. It is difficult to exaggerate the importance to Martian theorizing of the definite fact that Mars has oxygen and, therefore, vegetation.

A CERTAIN way along the path of evolution, Martian life shows evidence of having undergone a development like that on earth. What happened after that is a matter of deduction.

The known facts about Mars are the fruits of years of astronomical observation and study. The dark and light markings on its surface can be seen through a large telescope. The lighter ones are reddish or yellowish and usually are interpreted as being deserts. The darker areas are greenish or bluish in color and are universally ascribed to vegetation. Mars possesses

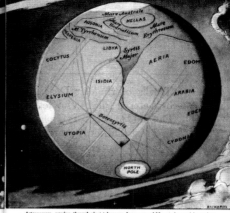

MAY, 1930   POPULAR SCIENCE MONTHLY

Astronomers, peering through giant telescopes, have mapped Mars and named its various features. Note the polar ice caps, which change with the coming of the winter or summer.

two white polar caps. Recent measurements of Martian temperatures by Dr. W. W. Coblentz and Dr. C. O. Lampland, at the Flagstaff Observatory, indicate that these are composed of snow and ice.

In the Martian autumn these caps increase and become whiter. In the planet's spring they shrink and often seem to be surrounded by wide rings of bluish or blackish material, which may be sheets of water or vegetation. Still more significant are the springtime changes in the planet's areas of supposed vegetation. Many of these darken in color. Others widen or lengthen. Often new dark areas appear where none had been visible during the Martian winter. Few astronomers now doubt that these dark areas represent some kind of vegetation.

SO FAR, everything runs strikingly parallel with evolution on earth. It is probable that it will be found to have run parallel farther still and that animal life on both planets, too, has been similar—for at least part of the evolutionary story. But during all the years of earnest and competent research not one clear sign of manlike life on Mars has been detected. Professor Lowell's famous Martian "canals," which for a long time were considered a probable sign of the intelligent direction of water, are now believed to be wide, shallow river valleys.

This lack of manlike life is precisely what a biologist would expect. Man and man's active mind are believed to be products of the Great Ice Age, for that time of stress and competition on earth is what is supposed to have turned mankind's anthropoid ancestors into men. The period of ice and cold over wide areas of the earth was caused, at least in part, by the elevation of continents and mountain ranges. On Mars, no mountain ranges exist, and it probably never had an Ice Age.

It is on these hypotheses that science bases its assumption that there is no human intelligence on Mars, and that animal life on the planet is still in the age of instinct. The thing to expect on Mars, then, is a fish life much like that on earth, the emergence of this fish life onto the land, and the evolution of these Martian land-fishes into reptilelike creatures. Finally, animals resembling the earth's present rodents like rats, squirrels, and

Big eyed, fast breathing beaverlike creatures may be the dominant animals on the cold and level planet Mars.

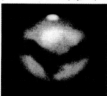

Scientists no longer believe there are man-made canals on Mars, and think the broad bands are the wide beds of wandering streams.

MARTIAN BEAVERS

In an article Thomas Elway wrote for the May 1930 issue of the magazine *Popular Science*, he expressed the belief that conditions on Mars would not be amenable to the evolution of human-like creatures. He did believe, however, that Mars had plant life. And, he reasoned, if there were plants there must be animals to eat them – although these animals would still be "in the age of instinct" and not intelligent.

Over the course of millennia, Elway suggested, evolution on Mars eventually produced a creature perfectly suited to live there. "Now," Elway wrote, "there is one creature on Earth for the development of whose counterpart the supposed Martian conditions would be ideal. That animal is the beaver. It is either land-living or water-living. It has a fur coat to protect it from the 100 degrees below zero of the Martian night." Elway urged his readers not to laugh at his giant beavers, with their enormous eyes and huge claws (for creating burrows deep in the Martian soil). After all, he said, "herds of beaver-creatures are at least a more reasonable idea than the familiar fictional one of manlike Martians digging artificial water channels with vast machines or the still more fantastical notion of octopus-like Martians sufficiently intelligent to plan the conquest of the Earth."

ABOVE: The title of this *Popular Scientific Monthly* article says it all. The frigid wastes of the Martian polar caps are the kind of environment that author Thomas Elway imagined a sturdy animal like the beaver was ideally suited for in his article.

ABOVE LEFT: The Martian landscape, as imagined beneath sheets of ice.

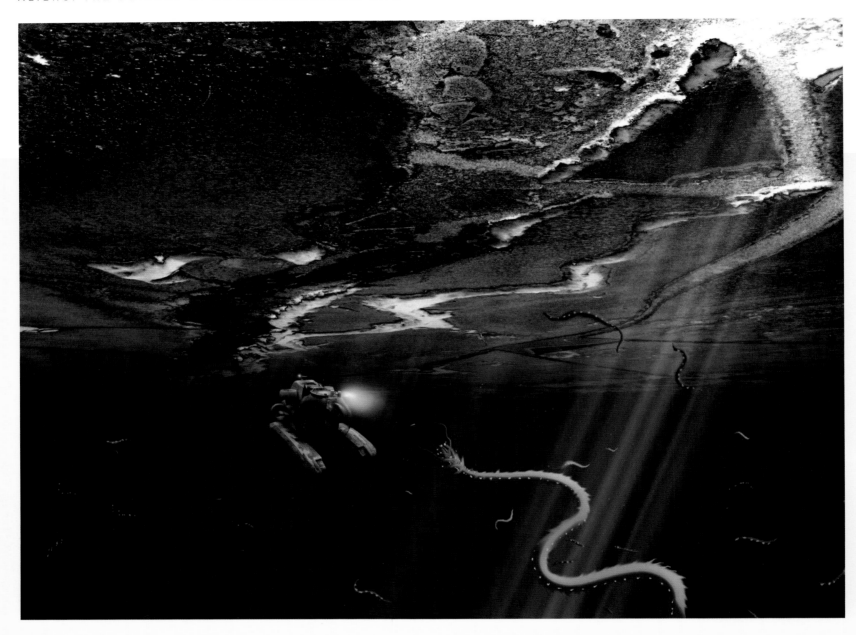

organic compounds. (Organic compounds are also responsible for the red, yellow and orange colours found in the clouds of Jupiter and Titan.)

Although Europa's subterranean ocean is hidden from the Sun, it still receives energy. This comes from Jupiter in the form of *tidal flexing*. As Europa orbits the giant planet, Jupiter's gravity causes the moon to flex. The effect is exactly like squeezing a rubber ball in your hand. When you squeeze it rapidly, it grows warm. In the same way, the flexing of Europa by Jupiter creates heat beneath the surface. This has kept the ocean from freezing and may

provide more than enough energy to make up for the lack of sunlight. The icy crust above the ocean also serves to protect the ocean by acting as a shield against Jupiter's powerful radiation.

What kind of life might exist in the deep, lightless seas of Europa? The most conservative scientists expect to find at least microorganisms similar to those found flourishing near undersea hydrothermal vents on Earth. But given that Europa's ocean may be at least 4 billion years old, there has been plenty of time and opportunity for the evolution of a larger and more vigorous biosphere. After all, more than just micro-

ABOVE: What kind of life might exist in the lightless depths of Europa's oceans? Like the depiction from this artist, we can only guess at the reality until submersible probes are able to penetrate the thick ice shell that surrounds the satellite.

ABOVE: In 2016 scientists determined the possible existence of geothermal vents in the seas that lay beneath the icy crust of Saturn's moon, Enceladus. As seen in this artist's recreation, these islands of warmth could be rich in nutrients and may be havens for various forms of life.

organisms live in or near the c370°C (700°F) waters that surround Earth's hydrothermal vents. Sea worms, shrimp and other complex life forms also thrive there. Even larger animals, such as fish-like creatures, may exist there if oxygen levels are sufficiently high. The lack of sunlight would hardly be an issue.

The abyssal depths of Earth's oceans are also lightless. The creatures that live there produce their own light . . . not in order to illuminate their environment, but to attract food and potential mates. So the oceans of Europa may also be alight with living neon signs.

Enceladus, the tiny – 804km (500-mile) wide – moon of Saturn, also possesses a worldwide ocean beneath its icy crust. Like Europa, this ocean is warmed by tidal flexing, which is caused by Saturn. Some of Europa's water escapes in the form of geysers, but in the case of Enceladus this occurs on a grand scale. Beautiful, feathery plumes of ice crystals spray hundreds of miles into the sky above the moon's surface. Analyses of these plumes have shown the presence of nitrogen and carbon dioxide along with sodium chloride (table salt), sodium carbonate and simple hydrocarbons such as methane, propane,

OPPOSITE: As seen in this artist's impression, methane rivers and hot springs dominate this landscape on Saturn's giant moon, Titan. One of the most Earth-like of the solar system's many worlds, it possesses all of the ingredients for the evolution of life in spite of its frigid temperatures.

ABOVE: Gas giants such as Jupiter, Saturn, Uranus and Neptune, which have no solid surfaces, may still be home to life. Artist Joel Hagen depicts the strange creatures, like living blimps, that may have evolved and exist within the clouds surrounding these worlds.

acetylene and formaldehyde. Although Enceladus' sea may be on the alkaline side, it still possesses a life-friendly chemistry. Recent evidence suggests that there may be hot hydrothermal vents on the floor of Enceladus' sea, similar to the undersea volcanic vents known as "black smokers" on Earth. These would provide ideal conditions for the evolution of life.

Titan, Saturn's largest moon, is a planet-sized body with an atmosphere denser than Earth's. The atmosphere is largely nitrogen while liquid methane fills lakes and rivers. The surface itself is covered in a dense layer of hydrocarbons. There is water there, but at -179°C (-290°F) it is as hard as rock and inaccessible to living organisms as we know them. Nevertheless, life could still exist on Titan. As long ago as the 1970s, astronomer Carl Sagan and chemist Bishun Khare showed that amino acids could easily form in Titan's atmosphere, and amino acids are the fundamental building blocks of proteins.

The discovery in July 2016, that Titan's atmosphere contains large amounts of hydrogen cyanide, was even more important. This chemical can serve as a precursor to amino acids and nucleic acids, which, in turn, can lead to the formation of proteins and DNA. Moreover, hydrogen cyanide molecules can link together to form polyimine, a molecule that supports prebiotic chemistry in the frigid temperatures found on Titan. Polyimine can also absorb a wide spectrum of light, including wavelengths capable of penetrating Titan's smog-like clouds.

Possibly the most extreme environment facing life in our solar system is the giant planet, Jupiter. Although its atmosphere possesses all of the necessary ingredients for the formation of organic compounds and the planet certainly has energy to spare, it is also an extremely hostile world. It is likely that Jupiter has no solid surface – its atmosphere grows denser and denser the farther one descends until it literally merges

"Billions of years and billions of biological accidents led to the diversity of life we see on Earth today."

Ron Miller

with the liquid hydrogen surrounding the planet. So, life would have to have evolved in the atmosphere itself, where it would be subjected to winds tearing across the planet at up to 1,609km (1,000 miles) per hour and temperatures as low as -168°C (-270°F). There would be no quiet sanctuaries – such as the tidal pools on the primordial Earth provided – where life could begin the long process of evolution.

But just supposing life *did* evolve on Jupiter – or any similar gas giant planet – what might it be like? With no solid surface on which it could ever rest, life forms might take the shape of giant living blimps. Atmospheric gases contained in vast envelopes might be heated by biological processes, turning the creatures into hot air balloons. This would allow them some control over the height at which they hover and, in turn, the direction in which they travel by choosing the layer of air in which the prevailing winds are travelling in the right direction – exactly as free balloonists do here on Earth. Creatures might resemble giant parachutes while others may be like Sagan's proposed aerial jellyfish navigating the warm updrafts or bat-like gliders with huge wings.

More Conditions for Life

The incredibly wide variety of environments found on our own planet alone has produced an equally bewildering array of life forms – a variety that has coined the word "biodiversity". One can only speculate on the directions life on other worlds may have taken even if it began with the same simple organic molecules.

So long as a world isn't as impossibly hostile as, say, our own solar system's Venus – where it is hot enough to melt tin and it rains sulphuric acid

– there is a possibility of life evolving. The only problem it faces is *time*. It took billions of years for life to evolve from primitive single-celled organisms to all of the diversity we see now. During that time, the environment has to remain reasonably stable. For example, if a planet has a very eccentric orbit, it might swing so close to its sun for part of the year that it is impossibly hot and so far away for another part that it is a frozen wasteland. Even though it may spend part of its year in the Goldilocks zone – where a planet is neither too far from its star nor too close for life to evolve, but where conditions are "just right" for liquid water to exist, sometimes also known as the habitable zone – it is never long enough for life to get a foothold. The Goldilocks zone is dependent on the star's luminosity – brighter, hotter stars have habitable zones with a larger radius; smaller, cooler stars have ones with a smaller radius. Although many searches for worlds like our own focus on Sun-like stars, it's also possible for smaller, cooler red dwarf

ABOVE: A shown in this diagram, the size and temperature of a star determine where its Goldilocks zone lies. A small, cool, red star (top) will have a habitable zone – shown in green – that lies close to the star. A large hot star (centre) will have a habitable zone that lies much farther away. A star like our Sun (bottom) will have a habitable zone that lies in a position midway between regions that are too cold and too hot to support life.

RIGHT: The existence of a large moon may be a significant factor in creating a stable environment for life to evolve. Here we see a depiction of the Earth shortly after the formation of the Moon 4 billion years ago. The surface of our planet is still a red-hot sea of lava, while a gigantic Moon looms in the sky.

stars to have a habitable zone – although any planets orbiting close enough would likely be tidally locked, with one hemisphere permanently lit and another always in darkness.

But even if a planet orbits solidly within its star's Goldilocks zone, there are still problems it must face. The axial tilt of a planet might be just as crucial as its distance from its star. It is Earth's tilt of 23 degrees that creates our seasons. It has not changed significantly over many millions of years, providing life the steady environment it needed to develop without interruption. Some scientists believe we can thank our Moon for this.

A rotating top will wobble as it spins. This wobble is called precession. In the case of a planet, this is largely due to the gravitational

pull of its star. The plane in which a planet orbits can precess as well. When the precessing axis of rotation and the precessing orbital plane get in sync with one another, the combination can cause the planet's wobble to become chaotic. This, in turn, can cause dramatic changes in the surface conditions on the planet.

However, the gravitational effect of our unusually large moon acts as a kind of stabilizer that has prevented this kind of catastrophic precession from occurring on Earth. Many scientists were of the opinion that without the stabilizing effect of a large moon, life would be unlikely to evolve on a planet. Since Earth possesses a huge moon mainly by chance, it might be unlikely that other Earth-like worlds in the universe would have had the same stroke of

good luck, thereby adding one more reason to doubt that life might be common.

The good news is that it seems that the effect of a large moon might be overrated. New studies have suggested that the influence of other planets in our solar system would suffice to keep our planet stable so that even without a moon, the tilt of Earth's axis would never vary much more than about 10 degrees – not nearly enough to cause any problems.

Given how likely it is for life to have already evolved on Mars, Europa, Enceladus and Titan in our own solar system, it would not be entirely unreasonable to suppose that the same

processes would be taking place on other worlds in other solar systems where conditions are even remotely similar. For life as we understand it – which is the only life we know how to look for – to exist requires only water, the right chemicals, a source of energy and a stable environment.

What Are the Chances?

Natural phenomena only go to show how much not only the survival of a species depends upon chance, but life itself. Entire families of animals that have existed for tens of millions of years can be wiped out almost overnight. Those that survive may be forced onto entirely new

ABOVE: Life on Earth could be extinguished as easily as it was created. Any number of events, from the explosion of a nearby supernova to the impact of an asteroid similar to the one that wiped out the dinosaurs, could reset the clock of evolution on this planet. Here an artist has imagined an asteroid impact occurring near a large, modern city.

"The almost endless chasm of the universe must be rife with living beings."

Dave Eicher (2016)

evolutionary tracks. There was nothing inevitable about the evolution of human beings into the dominant species on Earth. As we have seen, any one of thousands of random biological events – some large and some small – dictated the evolution of life on our planet. Had any one of these gone a different way, Earth might be swarming with intelligent dinosaurs or cockroaches instead of the creature that evolved from a tiny shrew-like animal. There was never any guarantee that the evolution of the earliest single-celled organisms on this planet would have taken the course it did.

All of this worked out for Earth: life survived the gauntlet. But given how fragile life is and how many hazards it has to run, if life exists on any of the exoplanets (those outside our solar system), the chances it resembles life here on Earth are almost incalculably small. There is also the grim possibility that intelligence per se is not necessarily a good thing for long-term survival. In the past, being able to outthink a predator or invent a useful hunting tool was beneficial. But did we over-evolve our brains? After all, while it's nice to have things like philosophy, science, literature, art and poetry, we could surely survive quite well without them. We certainly live longer and happier lives due to the benefits of modern medicine, but we would still live without them. Our species did so for millions of years.

We have, according to many scientists, entered an entirely new geological epoch: the Anthropocene. It began when human activities started to have a significant impact on the ecosystems and even the geology of our planet. Given this sort of power, granted to us by virtue of our intelligences, we have the ability to ignite the next Great Dying – the fuse of which may

already be lit, and evidenced by the extinction of species. It is entirely possible that the evolution of intelligence is counterproductive to the continued existence of life on a planet. It is certainly too soon to tell from our own example.

So, given the obstacle course life on Earth has had to run and the fact that intelligent life has existed here for only a couple of hundred millennia, what are the chances of finding similar life on the billions of other planets in the universe?

In 2016 astrobiologist Caleb Scharf and chemist Leroy Cronin suggested a way to deal with this question. To do this, they created the equation:

$$\langle N_{abiogenesis}(t) \rangle = N_b \cdot \frac{1}{\langle n_o \rangle} \cdot f_c \cdot P_a \cdot t$$

Where:

Nabiogenesis (t) = the probability of the origin of life (abiogenesis)

Nb = the number of potential building blocks

no = the average number of building blocks (such as molecules) on the planet

fc = fractional availability of building blocks for time

Pa = probability of an origin event per set of building blocks per unit time

t = the length of time in question

The equation shows that the probability of life beginning on a planet is connected to the presence and quantity of the basic building blocks of life, which do not necessarily have to be the same that led to life here on Earth. The formula also suggests that life might more easily form in solar systems consisting of multiple planets. Both of these results greatly increase the odds of finding living organisms on other worlds in the universe.

THE INTERPLANETARY ZOO

In 1951 the legendary science fiction illustrator Edd Cartier created a series of illustrations speculating on what the inhabitants of the different planets and moons of our solar system might look like. Most of Cartier's ideas were based on what knowledge was available at the time regarding the particular conditions on these distant worlds. Cartier was also not only one of the first artists to depict genuinely *alien* aliens, totally distinct from human life forms, but may have also been the first to suggest the idea of living balloons.

BELOW, LEFT TO RIGHT: Edd Cartier's visions of extraterrestrial life continue to fascinate us even today, with their imagination, scientific reflection and sheer strangeness. Here we have (from left to right) a creature from a planet orbiting Tau Ceti, an inhabitant of Mercury and a balloon-like Venusian.

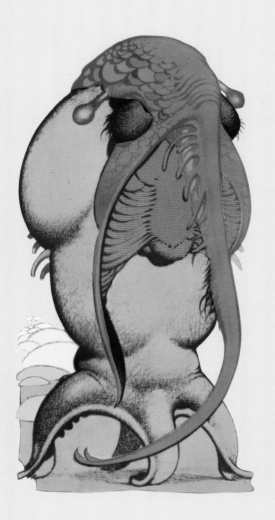  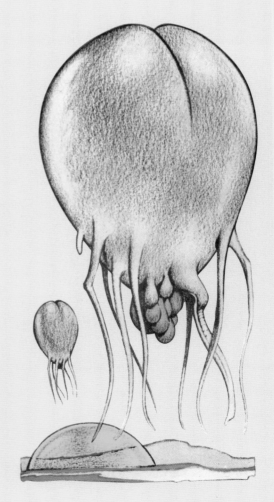

BELOW, CLOCKWISE FROM UPPER LEFT: One of the denizens of a planet orbiting the star Luyten-789-6; a massive animal from Jupiter; a Martian citizen; a native of Jupiter's moon, Io; a visitor from a planet of 61 Cygni; a plant-animal from one of the Messier galaxies; a creature from "beyond our galaxy"; and one of the diminutive beings that have no home world, but instead inhabit the space between the stars.

05

THE SEARCH FOR EARTH 2.0

OPPOSITE: Buried deep within the dense disk of dust and gas surrounding a young star, a newly formed planet is slowly accreting (gathering in matter) and growing in size. Although this is an artist's impression, the Hubble Space Telescope has observed numerous examples of this vast process taking place.

Both evidence and logic suggest that there should be planets around other stars. Still, at the end of the 1980s astronomers were still searching, and it wasn't until the 1990s that there was an unambiguous observation of one in reality. In 1983 the star Beta Pictoris was found to have a thin disk of dust surrounding it, similar to the disk that scientists believe once surrounded our own Sun during the early stages of the formation of our solar system. It was argued that if Beta Pictoris possessed such a disk, it seemed reasonable to assume that it might also have planets. Unfortunately, at that time the technology didn't exist to detect them. Or at least not planets that were recognizable as such.

Doppler shift measurements were able to show that at least some stars had one or more very large bodies orbiting them. But as they were many times bigger than Jupiter, our solar system's largest planet, it was thought that these were not true planets. Instead, they became known as brown dwarfs — objects much bigger than planets but not quite large enough for their gravity to ignite the nuclear fires and form stars. They would glow dimly like hot coals, hence their name. Although the planet hunters

failed to find true planets, the success in discovering brown dwarfs — "almost" planets — showed that they were on the right track and that their techniques simply needed more refinement and practice.

Pulsar Sign

Astronomer Alex Wolszczan was studying pulsars at the Arecibo Radio Telescope in Puerto Rico in 1990 when he made an astonishing discovery. Pulsars are tiny, ultra-dense, fast-spinning stars that emit a powerful beam of radio energy as they rotate at incredible speeds. The beam sweeps the sky like the beam of light from a lighthouse. Just as the light from a lighthouse seems to flash or blink every time the lighthouse lantern swings toward us, the pulsar's beam of radio energy seems to "pulse" every time it swings toward Earth. As this beam sweeps past Earth hundreds or even thousands of times a second, radio telescopes detect it as a series of clicks, as regular as the ticking of a precise clock.

But there was something irregular about pulsar PSR 1257+12. Sometimes a pulse came a minute fraction of a second late, while other times it came a fraction of a second early. This would be impossible unless the pulsar had

OPPOSITE: The first extrasolar planet to be discovered is the one orbiting the pulsar PSR 1257+12, seen here in an artist's recreation. The intense radiation that a pulsar produces makes it hard to imagine a more hostile environment and one less conducive to life.

RIGHT: The first planet found to be orbiting a Sun-like star was 51 Pegasi b. The planet itself would be unlikely to house life. A Jupiter-sized world orbiting twenty times closer to its star than Earth is from the Sun, it would have a surface temperature in excess of 1,000°C (1,800°F).

one or more planets orbiting it. As the planets swung around the pulsar, their gravity would pull it first toward Earth and then away. This could affect the timing of the pulses.

Wolszczan believed that this was exactly what was happening. In fact, he decided that there were two planets orbiting PSR 1257+12. One was about three times the mass of Earth and took about 98 days to circle the pulsar, and the other was about 3.5 times the size of Earth and orbited in just over 66 days. Both planets were roughly the same distance from the pulsar that Mercury is from the Sun.

This was the first evidence anyone had ever discovered of true planets around a star other than the Sun. But the worlds orbiting a pulsar would have little chance of life existing on them. Pulsars emit powerful X-rays which would be deadly. Although the discovery of these planets was exciting, it also raised some important questions. Was it only an accident that planets had formed around this particular pulsar? Worse yet, what if planets were more likely to form around pulsars than around stars like the Sun?

This would be bad news for the search for life since it would be unlikely for life to have even begun on such an inhospitable world, let alone evolve. So astronomers continued their search for a planet around a star similar to our own Sun.

The Planet Hunters

In the year 1995 Michel Mayor and Didier Queloz, astronomers with the Geneva Observatory in Switzerland, found a planet orbiting the star 51 Pegasi. They had been searching for the telltale wobble in stars that indicates the possible presence of planets. They had examined 150 stars before discovering what they had been looking for. The planet they found is about seven times larger than Earth, but still only about half the mass of Jupiter and therefore far too small to be a brown dwarf. It was a real planet. Unfortunately, the planet orbits very close to its sun (its year would be only four days long), nearly ten times closer than Mercury orbits the Sun. This is so close that its surface temperature is over 999°C (1,830°F), hot enough to bring rock to a red heat or to melt lead, tin or silver. Since

"Our sun enlightens the planets that belong to him; why may not every fixed star also have planets to which they give light?"

Bernard Le Bovier de Fontenelle (1686)

the planet is probably a nearly molten ball of rock and iron under seven times the surface gravity of Earth, the glowing landscape may flow slowly under its own weight like a glacier, so that mountains and craters are less permanent than a hole scooped in thick mud.

It would be just as impossible for life to exist on 51 Pegasi's planet as on the planets orbiting PSR 1257+12. But there is an important difference for consideration: 51 Pegasi is a star similar to the Sun rather than a bizarre object like a pulsar. One planet suggests there may be others, and these others may orbit far enough from the star for temperatures to be amenable to life. Although no other planets were found orbiting 51 Pegasi, astronomers were encouraged to continue searching similar stars for the telltale signs of planetary systems.

Just over a year later, in January 1996, astronomers Geoffrey Marcy and Paul Butler discovered planets orbiting the stars 70 Virginis and 47 Ursae Majoris. Both are stars similar to our Sun. The planet orbiting 70 Virginis, a star a little cooler and older than our Sun, may be only a brown dwarf. But the planet orbiting 47 Ursae Majoris has a mass of only about 2.5 times that of Jupiter. It orbits its star at a distance of about twice that of Earth from the Sun, a little farther than Mars is from the Sun, and takes about three years to make one orbit. It, too, may be cool enough to allow for the existence of liquid water on the planet.

2,000 Planets and Counting

In the two decades since the first exoplanets were discovered, nearly 2,000 have been found. Of all these new solar systems, 500 are multi-planet systems like our own. But most of these planets would not be habitable – at least not by life as we know it. In fact, most of these planets

OPPOSITE: Discovered in 1996, 47 Ursae Majoris b is an extrasolar planet. It is a frigid world larger than Jupiter and lies outside the habitable zone of its star. Here we see the ghostly planet from the surface of one of the moons it may possess.

ARMCHAIR PLANET HUNTERS

Through the initiative Planet Hunters (http://www.planethunters.org), ordinary citizens who have never even owned a telescope can participate in the scientific search for exoplanets. Created in 2010 by Zooniverse, an online platform encouraging collaborative research, Planet Hunters works with more than 300,000 volunteers worldwide. The project depends on the superior ability of the human eye to recognize visual patterns in the data being returned by NASA's Kepler Space Observatory,

patterns that might actually be overlooked by a computer. These patterns can indicate the presence of extrasolar planets.

In just the first two years that the programme had been operating, more than 40 candidate planets were found as well as a confirmed Neptune-like world. This planet – 5,000 light years from Earth – was the first ever found to be orbiting four stars. This planet has been dubbed "PH1" as the first to be discovered by the Planet Hunters programme.

ABOVE: The SETI@home project enables anyone with a computer and a little spare time to participate actively in the discovery of new extrasolar planets.

LEFT: Kepler-452b, discovered in 2015, was the first potentially rocky planet discovered orbiting within the habitable zone of a star similar to the Sun. A "super-Earth", the planet is 50 per cent larger than Earth, with a gravity twice as great.

OPPOSITE: Kepler-186f was the first Earth-sized planet found orbiting within the Goldilocks zone of its star. Although the star is a small, cool red dwarf, the planet orbits close enough that the temperature may allow for the presence of liquid water and consequently the possibility of life. Here an artist imagines a possible form that life may take. The plants are a very dark colour to better absorb energy from the dim star.

are so hostile that it would be unlikely for any sort of life to have evolved there. Indeed, some of these worlds are so bizarre that it's amazing that the planet itself exists.

Nevertheless, NASA's invaluable Kepler Space Observatory has found that 22 per cent of the stars it has observed have planets that are potentially like our own Earth. Since Kepler has examined only a very tiny fraction of the stars in our galaxy, scientists have extrapolated from this sample and have concluded that there may be more than 2 billion planets in our galaxy capable of supporting life. Such a planet would be "Earth-like", a potential candidate for that holy grail: Earth 2.0. (An "Earth-like" planet is one that is small enough to be rocky – rather than being a gas giant like Jupiter or Saturn – and orbits its star within the "habitable zone".)

In 2014 Kepler confirmed the first near-Earth-size planet in the "habitable zone" around a sun very similar to our star. Kepler-186 is a five-planet star system about 500 light years from Earth in the direction of the Cygnus constellation. The five planets of Kepler-186 orbit a red dwarf star that is half the size and mass of our Sun. Red dwarf stars make up 70 per cent of the

stars in our galaxy and hundreds of them have been discovered to have one or more planets orbiting them. Of the five planets belonging to Kepler-186, Kepler-186f, the outermost, orbits the star every 130 days and receives only one-third of the energy from its star that Earth gets from the Sun. This places it nearer the outer edge of the habitable zone, like Mars in our solar system. Although Kepler-186f is closer to its star than Earth is to the Sun, its star is smaller and dimmer. High noon on Kepler-186f would be only as bright as an hour before sunset on Earth.

One of the most exciting discoveries in the hunt for Earth 2.0 occurred in August 2016, when a planet was found orbiting the star Proxima Centauri. Dubbed Proxima Centauri b, it is a rocky world only slightly larger than Earth. It orbits very near its star – at a distance of just 0.05 AU – and it has a "year" of only 11 days. But since Proxima Centauri is a small, dim, red dwarf star, the planet receives a little less light and heat than the Earth does from the Sun. Nevertheless, it falls well within the Goldilocks zone of Proxima Centauri, making conditions there "just right" for the potential evolution of life. Though whether this has happened – or is even possible – is yet to

"A time would come when men should be able to stretch out their eyes. . . . They should see planets like our Earth."

Sir Christopher Wren (1657)

be determined. One stumbling block is that the star is an X-ray flare star, which means that the intense radiation that bombards the planet may prove to have been an impediment to the development of life.

What is particularly exciting to the scientific community is that Proxima Centauri is the nearest star to Earth (not counting the Sun, of course). Until the discovery of Proxima Centauri b, the nearest habitable planet was more than 20 light years from Earth. This means that even travelling at the speed of light, a one-way trip would take 20 years. The Apollo spacecraft that carried humans to the Moon would take half a million years to make the journey. Since the new planet is only 4.5 light years away it might be possible for a spacecraft to make the journey within a lifetime, given the technological

advancements in propulsion that have occurred in the past 50 years.

Better than Earth

Our own Earth has always been the gold standard for planets capable of evolving life. It seems fortunate enough to have all the right ingredients in all the right proportions. It has liquid water in abundance, a stable environment and it sits right in the middle of the Sun's Goldilocks zone. Could it get any better than this? A growing number of scientists think so.

Instead of looking at our planet as the model for the ideal life-bearing world, they instead asked themselves: "If one were to build a planet from scratch with the sole goal in mind of making it as hospitable to life as possible, what would that planet be like?" The answer turned out to be

ABOVE: In 2016 an Earth-sized planet was discovered orbiting within the habitable zone of Proxima Centauri, the star nearest our solar system. Since it is only 4.5 light years away, it is the first exoplanet discovered that humans may be able to explore in the future. An artist has imagined what the surface of that world might look like, with its sky dominated by a star appearing seven times larger in the sky than our own Sun. But since Proxima Centauri is a dimmer, cooler star than the Sun, the surface of the planet receives no more heat than the Earth does from its Sun.

"In spite of the opinions of certain narrow-minded people, who would shut up the human race upon this globe... we shall one day travel to the moon, the planets, and the stars."

Jules Verne (1865)

ABOVE: There may be planets in space that are even more conducive to the evolution of life than our own Earth. These "superhabitable" worlds could offer conditions better suited for life to emerge, evolve and flourish. Some of these characteristics might include more land area and smaller, shallower seas.

a surprising one: it would be a world different from our own.

The ideal "Earth" would be slightly larger, about two or three times its present radius. The larger core would generate a more powerful magnetic field, providing better protection from solar radiation. There would also be more

volcanoes to pump carbon dioxide and water vapour into the atmosphere. The greater gravitation of the larger world would mean that the atmosphere itself would be denser. Plus, a bigger planet would have that much more surface area – 13 times more for a planet with just 3 times the radius of Earth – for life to take hold and find a habitat.

A "superhabitable Earth" doesn't even need to orbit within the Goldilocks zone of its star, either. Tidal heating – like that which occurs on Europa, Io and Enceladus – could provide warmth for a world lying outside the Goldilocks zone. It is easy to imagine a planet as large as ours or even larger orbiting a super-Jupiter like a moon. In short, a planet capable of evolving and supporting life doesn't need to closely resemble Earth at all.

THE EARTH SIMILARITY INDEX

How close a planet might be to Earth 2.0 can be judged by its score on the Earth Similarity Index (ESI). This is a measure, devised by the University of Puerto Rico at Arecibo, of how similar to Earth an exoplanet might be. It ranges between zero (no similarity) to one (identical to Earth). The search for Earth-like planets is synonymous with the search for planets with ESI values closest to one. The ESI is neither a precise measurement nor a direct indication of a planet's habitability, but is the product of comparing as many different parameters as possible. Some of these factors include the planet's radius, density, escape velocity and surface temperature. The closer a planet gets to 1.0, the more likely it is to be a rocky world. The more different qualities included in the calculation, the closer the resulting number will be to suggesting how Earth-like a world might be in reality.

In our own solar system, Earth has an ESI of 1.0, of course. Mars rates at 0.7 and Venus at 0.44, but Jupiter scores only 0.3 and Pluto 0.08. Kepler-186f scores at 0.64 and Kepler-452b has an ESI of 0.83. But it is only the sixth most Earth-like planet yet discovered. Kepler-438b is the leader so far with an ESI of 0.88. There is not enough known yet about Proxima Centauri b to assign any realistic number to it – but that will come as more research is carried out.

RIGHT: A sufficiently advanced civilization may be able to utilize all of the energy produced by the radiation from its star by surrounding it with enormous, energy-absorbing structures, such as those seen surrounding the stars in this artist's interpretation. Such energy-absorbing shells are called "Dyson Spheres" in honour of Freeman Dyson, who first suggested them.

OPPOSITE: An artist has imagined what an Earth-like alien planet might look like. Here we see a trinary planet system: three Earth-like worlds orbiting one another and together orbiting their star.

Is There Intelligent Life Elsewhere? How Would We Recognize It?

Astronomers around the world were puzzled by the characteristics and behaviour exhibited by the star KIC 8462852. For several years they had observed its brightness fluctuating dramatically, periodically dropping in intensity by up to 22 per cent. A change in brightness of that order seemed too great to be caused by an orbiting planet, while the effect of a planet-forming disk didn't seem plausible either.

KIC 8462852 is a mature star and if it had any planets they would have already formed long ago. A few more imaginative astronomers suggested, however, that the dimming might be caused by some type of orbiting alien megastructure. The idea is that the star might be surrounded by a Dyson Sphere.

This was proposed as a thought experiment by physicist and astronomer Freeman J Dyson in 1960. Earth intercepts only a tiny fraction of all the energy the Sun broadcasts into the universe. He suggested that a civilization that was advanced enough could collect all of the energy from a star by surrounding it with a kind of shell. This would not be a solid structure (which would actually be impossible), but would instead take the form of a ring or vast cloud of solar collectors or habitats. These might eventually completely encase the star so that no energy at all is lost to space. (This idea was actually first mentioned by Olaf Stapledon in his 1937 science fiction novel *Star Maker*, which may have inspired Dyson.) Dyson realized that such a structure would have a distinctive infrared fingerprint and suggested that astronomers looking for signs of extra-terrestrial intelligence might want to search for signs of Dyson Spheres.

While the existence of such a megastructure is a long shot, some astronomers – beginning in 2012 – have aimed radio telescopes at KIC 8462852 in the hope of discovering the telltale signature of a Dyson Sphere. Unfortunately for the hopes of finding an alien megastructure, however, the result of closer studies suggests that the weird effects are most likely the result of the star's strange shape and the presence of at least two transiting planets.

FAR LEFT: Percival Lowell's theories about the possibility of life on Mars had so captured the public imagination that composers were inspired to write music based on the idea of communicating with the red planet. This march and two-step was composed by Raymond Taylor and E T Paull in 1901.

LEFT: In the latter half of the 19th century, several scientists suggested that it might be possible to communicate with Mars by means of signals reflected by enormous mirrors. The best known of these schemes was that outlined by the French inventor Charles Cros, who published in 1869 his plans to signal the planet with a giant mirror device.

Listening for ET: From Tesla to SETI

As we have seen, as soon as humans learned that there were worlds other than their own, they began wondering what sort of life might live there and how to discover it. The question became: even if there were intelligent beings on the planets, how could you communicate with them?

The great German mathematician Carl Friedrich Gauss (1777–1855) thought it might be possible to announce to the universe that intelligent life existed on Earth. His plan was to create an enormous diagram illustrating the Pythagorean Theorem. This would be done by planting vast fields of wheat or rye in geometric shapes in the Siberian tundra. In the early 1800s, astronomer Joseph Johann von Littrow suggested digging trenches in the Sahara Desert, filling them with oil and igniting it. The blazing message would be visible from far into space.

Almost as soon as Percival Lowell announced his "discovery" of canals on Mars, and his theory that they must have been constructed by sentient creatures, imaginative plans were suggested as to how to contact the Martians. The French poet and inventor Charles Cros (1842–88) wanted to build a giant mirror on Earth. It would be used to

focus sunlight on Mars, where a message could be burned into a Martian desert, hopefully spelling out, "Sorry about that".

Forty years later the German biologist Wilhelm Bölsche came up with a highly original idea that was inspired by the theory of panspermia, which had been recently revived by the Swedish scientist Svante Arrhenius. While Arrhenius' idea was to explain how the seeds of life might spread through interplanetary and even interstellar space, Bölsche thought this might provide a means of communicating with other worlds. He proposed that geometric diagrams be microengraved on grains of dust. These would then be cast into space to be borne away by the solar wind.

By the end of the 19th century, some scientists speculated that the newly invented radio might be a potential method of communicating with other worlds. Both inventor Nikola Tesla and his arch-rival Thomas Edison thought they had detected radio signals from space. In the case of Tesla, his signals were from terrestrial broadcasts while Edison had discovered radio emissions from the Sun. So certain was it that planet Mars was inhabited that when the wealthy Parisian Clara

"Humans first detected our galaxy's radio emissions more than a century ago."

Ron Miller

Gouget Guzman established a prize in 1900 for the first person to communicate with another planet, she specifically omitted Mars because she deemed that it would be too easy.

The invention of the radio telescope in the 1930s, along with the realization that radio signals could be detected at interstellar distances, eventually caused the idea of interplanetary communication to be turned around. Instead of trying to send signals to other worlds, it would be infinitely easier and cheaper to listen for signals from other worlds.

Astronomer Frank Drake created "Project Ozma" in 1960, the first attempt to detect radio signals from extraterrestrial civilizations. Using the radio telescope at Green Bank, West Virginia, he scanned selected stars, looking for anomalous signals. His search proved fruitless, but it led eventually to the creation of several new SETI

(The Search for Extraterrestrial Intelligence) projects around the globe. To coordinate some of this research, the SETI Institute was created in 1984. Today it has invested more than a quarter of a billion dollars in research and employs 140 people administering more than 100 different projects (including studies related to planetary exploration and astrobiology, in addition to SETI). By using SETI@Home, created by the University of California, anyone with a computer, anywhere in the world, can take part in the search for extraterrestrial civilizations.

Although some form of SETI has been listening to the skies for decades, it has yet to discover unambiguous evidence of an extraterrestrial civilization. There have been tantalizing hints, beginning with the famous "Wow!" signal of 1977 (see feature on p.111). The most recent was found in August 2016: a strong signal coming

RIGHT: This diagram, published in 1951, anticipated the famous Voyager plaque by two decades. Its inventor hoped to be able to communicate basic ideas about Earth and its technology to its neighbouring planet, Mars.

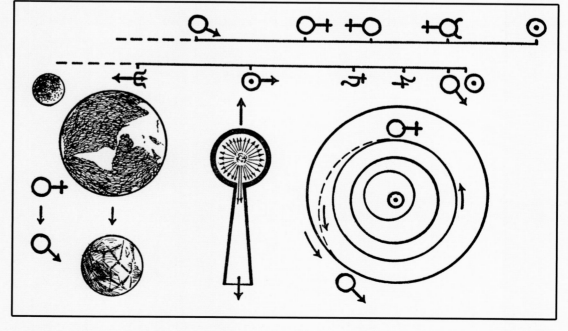

"If someday such life is discovered, will it not be in the image of man? This is what I believe."

Win Brooks (1948)

from the direction of HD164595, a star similar to the Sun that is 95 light years away. It is already known to possess a Jupiter-sized planet, which suggests there may be others. While the artificial nature of the signal is still open to question, scientists have placed the star under permanent monitoring.

In the meantime, the desire to announce our presence to the universe had not died. When it was realized that both the Pioneer and Voyager spacecraft would eventually leave our solar system entirely, messages were attached. In the case of the two Pioneer probes, this took the form of an engraved, gold-plated plaque.

Conceived by astronomers Frank Drake and Carl Sagan, the artwork – designed by Sagan's then-wife Linda Salzman – depicted male and female human beings along with a diagram of Earth's location in space. Launched in 1972 and 1973, by 2010 the spacecraft were deep into interstellar space. In the case of Pioneer 10, it had reached a point 100 times farther from the Sun than Earth. Even so, Pioneer 10 still has 2,700 times farther than that to go before reaching the nearest star, nearly 108,000 years from now. We do not know what they will eventually find, but is it possible that we could find extraterrestrial life even closer to home?

ABOVE LEFT: Voyager 1, which left the solar system in 2012, will pass near the star AC +79 3888 in 40,000 years. It carries a golden record bearing images and sounds from planet Earth.

ABOVE RIGHT: In 1974, Frank Drake, with advice from Carl Sagan and other scientists, devised a message sent into space from the Arecibo radio telescope. When decoded, the message provides such information as the numbers from 1 to 10, the atomic numbers of the elements that make up DNA and the location of Earth. The message is aimed at the star cluster M13, taking 25,000 years to get there.

OPPOSITE: Both Pioneer 10 and Pioneer 11 left the solar system in 1983. Pioneer 10 will arrive in the vicinity of the star Aldebaran in about 2 million years, while Pioneer 11 will pass close to the nearest star in the constellation Aquila in about 4 million years. Both carry information plaques designed by Frank Drake, Carl Sagan and Linda Salzman Sagan, the latter of whom created the famous drawing.

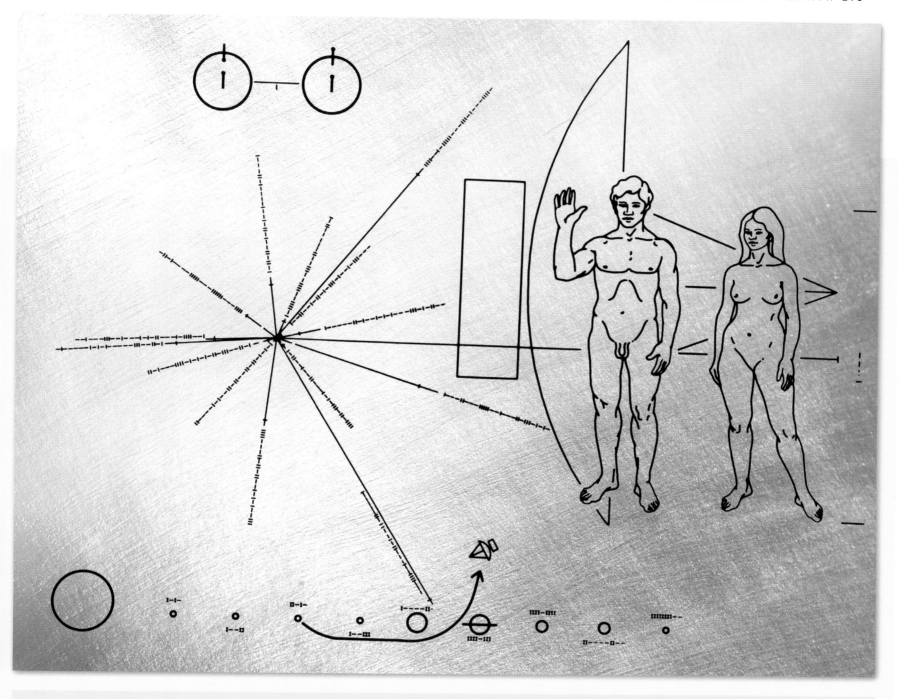

THE "WOW!" SIGNAL

On 15 August 1977, the astronomer Jerry Ehman was looking over the extensive computer printouts of the "Big Ear" radio telescope at Ohio State University. What he found startled him. A sudden, unexpected peak in the otherwise undistinguished noise. An excited Ehman circled the signal, scribbling "Wow!" next to it. Was it the long-awaited sign of an extraterrestrial intelligence? For the next month astronomers scanned and rescanned that part of the sky more than 100 times . . . and the signal was never repeated. Some suggested that it might have been a transmission from a classified military satellite or perhaps an ordinary radio signal from Earth bouncing off space debris. In either case the signal would surely have been repeated as the object orbited Earth, but the "Wow!" signal has never appeared again.

AN EARTH-LIKE LANDSCAPE UNDER THE LIGHT OF DIFFERENT SUNS

Here we see imaginative reconstructions of the same Earth-like planet in the habitable zone of six different Sun-like stars. Not only are the conditions for life itself determined by the amount of heat emitted by each star, even the colour of plants (where they exist) is affected by the different spectrums. Plants growing in the dim light of cool red stars, for example, are dark blue or purple, enabling them to absorb energy more efficiently.

THIS PAGE & OPPOSITE: The same landscape would look very different under various Sun-like stars, as these images by author Ron Miller show. On the facing page is what we might find on a world orbiting an M1 star. The same world, beginning clockwise from the upper left on this page, orbiting a G-type star would be familiar-looking since our own Sun is this kind of star. We then see this world under a hot F-type star, an M5 star, a red dwarf K-type star and finally an M3 star, which radiates so intensely that life would be safest underwater.

03

"Contact with alien life would shake the foundations of human society — it would be the biggest discovery of all time."

Dave Eicher (2016)

ALIENS

AMONG

US

06

THE COMING
OF THE SAUCERS

The story of aliens from other worlds and the story of UFOs and flying saucers overlap, but are not necessarily always the same thing. For the first decade or two of the flying saucer phenomenon, it was tacitly assumed that whatever they were, they were probably visitors from outer space. However, today while what has come to be known as the "Extraterrestrial

Hypothesis" is still dominant, there are many other theories out there. For example, there are those who adhere to the idea that UFOs are psychic phenomena. This theory is championed by one of the first experts in UFOs, Jacques Vallée (the inspiration for the character Lacomb in the movie, *Close Encounters of the Third Kind*) who proposes that what we see as "flying saucers" are really nothing more than images projected into our minds by some unknown but powerful force.

Another camp holds that all UFO sightings are either illusions, misidentified natural phenomena or outright hoaxes. There is much to be said for their case. Indeed, the only two formal studies of the UFO phenomenon – *Project Bluebook* (1952) and the *Condon Report* (1968) – concluded that all but a small fraction of UFO reports could be easily explained. Needless to say, many jumped on that unexplained remainder and assumed that it represented cases that could *not* be explained. But unexplained and unexplainable are two very different things.

Yet another popular theory is that UFOs are actually living creatures. This is an idea that predates the modern "flying saucer" by many decades. Sir Arthur Conan Doyle – the creator

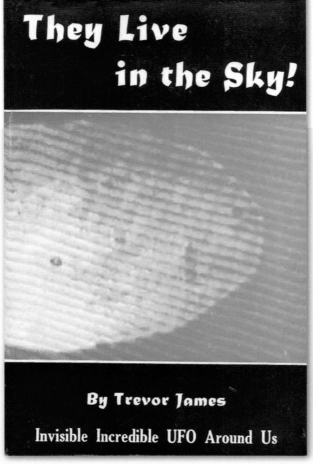

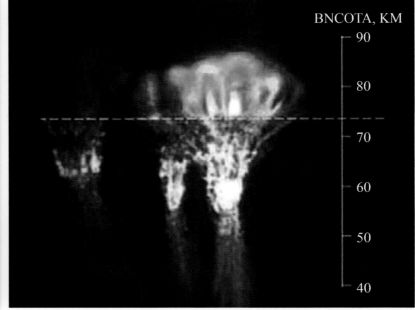

of Sherlock Holmes – published "The Horror of the Heights" in 1913. This short story tells of an aviator trying to break the contemporary altitude record of 9,144m (30,000ft) while at the same time solving the mystery of the strange deaths of earlier high-altitude seekers, one of whom was discovered with his head missing. The pilot believes the answer lies in "jungles of the upper air" in which entire ecosystems exist, inhabited by huge, gelatinous, semi-solid creatures that resemble enormous jellyfish-like creatures. He is tragically proven correct.

In his book *Lo!* (1931), that eminent and indefatigable researcher in the bizarre, Charles Fort, wrote that: "Unknown, luminous things, or beings, have often been seen, sometimes close to this earth, and sometimes high in the sky. It may be that some of them were living things that occasionally come from somewhere else." In recent times, there have been several

proponents of the "living UFO" theory. Trevor James Constable was one of the most active and vocal. His twenty years of research resulted in several books, including *They Live in the Sky* (1958) and *Sky Creatures: Living UFOs* (1978). Noted cryptozoologist Dr Karl Shuker is probably the best known among those who are carrying Constable's torch. As unlikely as it is that creatures hundreds or even thousands of feet across could fill Earth's stratosphere, the possibility that vast "sky jellyfish" exist on giant gas planets similar to Jupiter and Saturn is taken very seriously by many.

Some UFOs may have an electrical origin, similar to little-understood phenomena such as "earthquake lights" and ball lightning. This has been suggested as an explanation for the infamous "foo fighters" of World War II – glowing spheres of light that seemed to follow aircraft in their flight.

ABOVE LEFT: Trevor James Constable's book *They Live in the Sky!* (1958) was one of the first to propose seriously that UFOs might be living creatures.

ABOVE RIGHT: It is possible that the idea of "sky jellyfish" may have been inspired by the sighting of "sprites", an unusual electrical phenomenon that occurs in the upper atmosphere and which has been only rarely photographed, as in this image acquired in 2012.

OPPOSITE: Arthur Conan Doyle's short story "The Horror of the Heights" (1913) described an encounter with creatures living far above Earth. It was probably the first time that this idea had ever been suggested. This illustration by W R S Stott is from the original publication.

"The body was at least 100 feet long . . . it turned quickly and disappeared in the direction of San Francisco."

Case Gilson (1896)

Finally, there is the idea that UFOs have an entirely terrestrial origin. In fact, this was the original explanation for most of the early UFO observations, such as the aforementioned foo fighters and the "ghost rockets" of World War II and the years following. Since then UFOs have been attributed to everything from clandestine US and Soviet experiments to secret Nazi aircraft to refugees from Atlantis living in the centre of Earth. A subset of this theory is that of the *ultraterrestrial*. This is based on the idea that there is another intelligent species on Earth, a species that evolved along with *Homo sapiens*, but on an entirely different track. They are so far ahead of us in every way that we are barely even able to recognize them. All of the anomalous creatures of folklore, legend and the Bible – gods, angels, fairies, leprechauns, elves, dragons, monsters and so forth – are actually nothing but manifestations of ultraterrestrials mis-

remembered and misdescribed as they became part of our mythic and religious culture. The ultraterrestrials have their motives for what they do, but they are too far beyond anything humans could possibly understand or comprehend. A basic premise of this ambitious theory is that the creatures and events of folklore and myth were not the inspiration for modern UFOs, aliens, abductions, etc., but are in reality factual reports of exactly the same things and events that we experience today.

The Great 1890s UFO Flap

For a brief period during the late 1890s, a wave of UFO sightings swept across the United States. Dozens of strange objects were seen sailing through the skies from coast to coast. For example, in November 1896 an object was reported in the night sky over Sacramento, California. It was described as a light with a dark

ABOVE LEFT: The typical description of a UFO seen in the 1890s closely resembles the many airships – both real and imaginary – being depicted in magazines and newspapers across the country. The cigar-shaped body and propellers certainly seem to belong to something more terrestrial in origin than alien.

ABOVE RIGHT: This UFO was sighted in 1897 and resembles to a remarkable degree the airships then being experimented with by inventors across the world, particularly with respect to the cigar-shaped gas bag and the underslung gondola for passengers.

"I saw a disk up in the air,
A silver disk that wasn't there.
Two more weren't there again today –
Oh how I wish they'd go away."

Men's room graffiti, White Sands missile range, New Mexico (1967)

RIGHT: Jules Verne's 1886 novel *Robur the Conqueror* illustrated by Léon Benett, opens with a description of UFO encounters that sound remarkably modern, ranging from strange lights in the sky to mysterious sounds.

body of some kind above it. It was seen a second time about a week later. Similar reports, usually describing only the light, came from other cities in northern California. Reported movement of the light indicated a slow motion. The dark body seen above it was variously held to be "cigar-shaped", "egg-shaped" or "barrel-shaped". Other reports from the 1890s wave describe a fast-moving cigar-shaped object which glowed and made small explosions, similar to what might be expected from a gasoline engine. Another emitted coloured rays of light. An unknown body – a "luminous ball of fire" – circled a mountain in Canada before speeding away. The ship described in an 1896 close encounter in California was characterized as "cigar-shaped". An object seen in Kansas in 1897 looked like a 9m (30ft) canoe with a searchlight. In 1897 a strange craft allegedly hovered above a Kansas farmer's cow lot. It was a 91m (300ft) long cigar with a carriage underneath and, beneath that, a 9m (30ft) revolving turbine wheel. Windows lined the sides of the gondola and the interior was brightly lit. Another sighting involved a cigar-shaped object with four wings, a searchlight and fan-like wheels. In a different report, the motion of a light in the night sky was said to suggest "the flapping of wings". Another farmer reported an airship with flapping wings, while others ascribed hissing sounds to the ships they saw. There were many such reports, all with more or less similar descriptions.

In 1886 Jules Verne published his thirty-first novel, *Robur-le-Conquérant* (*Robur the Conqueror*). As were most of the novels he had written to this time, *Robur* was not only translated and published in English, it was pirated in editions all over the world, especially in the United States, where copying Verne's work had become something of a cottage industry. The French author was the inspiration for a series of wildly popular dime novels devoted to the adventures of a young inventor named Frank Reade, Jr. Most of these stories featured incredible flying machines resembling those in Verne's novel.

Robur opens with a detailed description of what today would be a UFO flap of epic

"I can assure you that, given they exist, these flying saucers are made by no power on this Earth."

President Harry S Truman (1950)

OPPOSITE: This illustration published at the end of the 19th century depicts only a handful of the fanciful airships being designed and experimented with by inventors across the globe. Many of these bear a close resemblance to the UFOs being reported at the same time.

proportions: mysterious objects have been observed in the sky all over the world, in Europe, Asia and America. Strange lights and sounds are seen and heard by thousands of people. Newspapers on every continent discuss, report and debate the phenomenon: "recording things . . . false and true, alarming and tranquilizing their readers as the sale required and almost driving ordinary people mad." Astronomers have no answer to the mystery.

Many of the reports are of aerial flashes of light. During the day the phenomenon manifests itself as "a small cloud or vapour". On one night in the centre of the aurora borealis there is seen the silhouette of some enormous, unknown structure "showering off from its body certain corpuscles which exploded like bombs". If it is indeed the same object being seen consecutively by different observers in the same night, then its speed must be unprecedented. At the same time that one group of astronomers is "explaining" the sightings as nothing more than optical and acoustical illusions or misinterpreted meteors, another group is hotly arguing for the existence of some unknown flying object in the atmosphere. The plot develops to reveal that the mysterious object is indeed a giant flying machine, invented by the novel's anti-hero Robur. He kidnaps three sceptics and takes them on a round-the-world flight (the first UFO abduction?).

The coming heavier-than-air conquest of the skies was one of the most popular topics of the late 19th century. Scarcely a week or a month would go by without some popular magazine publishing plans, drawings or photographs of some new flying machine. The questions scientists keep coming back to are: why do the

mysterious airships of the 1890s so uncannily resemble the airships imagined by Verne and his contemporaries? Why does their sudden appearance coincide with the proliferation of the "scientific romance" in which such machines are described in detail? Why does the appearance of these UFOs match the descriptions of actual airships being built or designed at the same time? Why were so many of the events described in Verne's novel duplicated in accounts of the mysterious flying machines?

During this brief period in the 1890s, UFOs are described as looking specifically like either real-life airships or those from popular fiction. The descriptions of the inhabitants of these machines, when they made an appearance, are completely prosaic. They speak colloquial English, wear contemporary clothing (one aeronaut was seen sitting on his flying machine, fishing, wearing a "checked hunting suit"!) and have normal human needs. Even the needs of the airships themselves are ordinary. They seem to have been in constant need of repair, oil, tools, fuel and water.

For thousands of years, people have been seeing aerial phenomena that they cannot explain. In the early history of mankind, these strange things were seen as manifestations of the gods: they looked like angels, devils, chariots or the gods themselves. In the latter part of the 19th century strange things in the sky purported to resemble the aerial craft then so much in the news: they looked to people like dirigibles and flying machines. In this century, whatever the unexplained objects in the sky really are, they are described almost invariably as *spaceships*.

Of course there are a great number of exceptions to this generalization, but it does

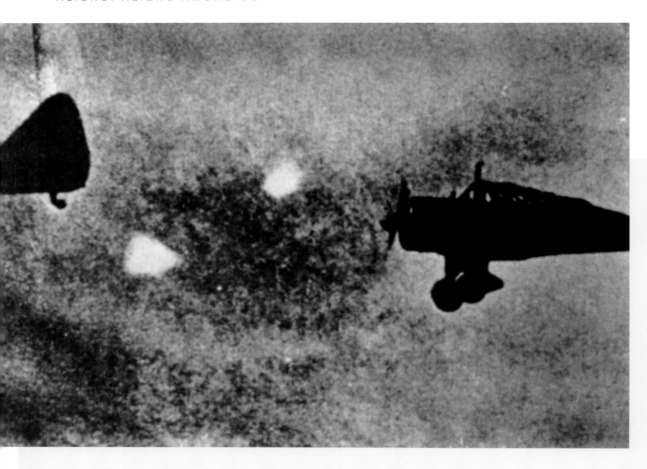

ABOVE: The mysterious foo fighters of World War II – seen here in a rare photograph – may in reality have been an electrical phenomenon similar to ball lightning or St Elmo's fire.

seem true that to a large degree UFOs typically look like what the percipient *expects* them to look like. In the religious and warlike atmosphere of biblical and medieval times, anything unfamiliar in the sky had to be, almost by definition, a sign from either God or the devil – and it was described that way. The natural phenomenon or "wheel", that so astonished Ezekiel (*see* p.200) is a perfect example. Bombarded from all sides by heavier-than-air ships – fictional and semi-fictional – the 19th century people described the strange things they saw in the sky as looking like the very aircraft they *expected* them to be.

To how great a degree are the descriptions of UFOs being coloured and shaped by preconceptions? If it is not unreasonable to suppose that these are the *same* types of phenomena that have been observed through the ages (whatever these phenomena may be), perhaps it is something a little more amorphous and vague than most observers might believe. How many ordinary but unfamiliar natural

phenomena were given wheels, wings, propellers or disc-shapes by the expectations of the observer, who believed these were the features it *must* have? This may be the most important lesson to be learned from the 1890s UFO scare.

Foo Fighters to UFOs

During World War II, fighter pilots reported seeing small spheres or disks of light that seemed to chase around their planes with apparent sentience. These apparitions were dubbed "foo fighters". Some have suggested that this was inspired by a phrase in the popular comic strip "Smokey Stover", by Bill Holman: "Where there's foo there's fire". More likely, it is a corruption of "feu", the French word for "fire".

Apparently first reported in 1944 and 1945, they were seen by pilots flying on all sides: American, British, German and Japanese. Each thought the foo fighters were secret weapons belonging to the other side. Shortly after the war this idea evolved into the "ghost rockets" reported

ABOVE: The mysterious foo fighters of World War II – seen here in a rare photograph – may in reality have been an electrical phenomenon similar to ball lightning or St Elmo's fire.

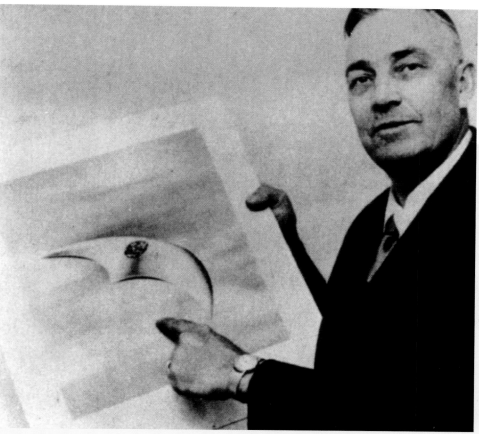

ABOVE LEFT: An interpretation by artist Ludek Pesek of Kenneth Arnold's sighting. The objects were described as seeming to be made of polished metal as they flew in formation against the background of mountains.

ABOVE RIGHT: Kenneth Arnold points to a drawing of one of the UFOs he saw in 1947. Arnold did not describe them as being shaped like saucers, but rather as crescents that moved in a motion resembling saucers being skimmed across a pond.

first in Sweden, then from locations as diverse as Denmark, Spain, Greece, French Morocco and Turkey. There were even a few reports from the United States. The ghost rockets were usually described as being metallic, cigar-shaped objects often equipped with fins. By the end of 1946 nearly a thousand had been "sighted" before the reports gradually stopped coming in.

The foo fighters were generally attributed to freak electrical or optical phenomena – or even outright hallucinations. (One foo fighter observed by the crew of a B-29 flying in the Pacific proved to be nothing more than the planet Venus in the early pre-dawn sky.) However, the ghost rockets may – at least in part – have been exactly what they looked like. A Swedish commission set up to investigate the sightings thought they might be V-1 "buzz bombs" launched by the Soviet Union. It is certainly true that both the Soviets and Americans were then actively developing the rocket technology each country had inherited from the Germans.

Whatever explanations were invented for these phenomena, they all had one thing in common: no one suggested that either the foo fighters or the ghost rockets came from outer space. There were some odd reports – fewer than two dozen – of unusual flying objects between the summers of 1946 and 1947, but no one paid them much attention. On 24 June 1947, all that changed forever. A businessman, ex-deputy Federal marshal and pilot named Kenneth Arnold was flying his low-winged private plane near Mt Rainier, Washington state, when he noticed a bright flash to his left. He was surprised to see "a chain of nine peculiar-looking aircraft flying from north to south at approximately 9,500 feet [2,895m] elevation and going, seemingly, in a definite direction of about 170 degrees." Every few seconds two or three of them change direction slightly. When they did, there would be a flash of sunlight like the one that first drew his attention.

By comparing them to the landscape they were passing through, he estimated the line of objects

"UFOs: The reliable cases are uninteresting and in the interesting cases are unreliable."

Carl Sagan (1975)

to be about 8km (5 miles) long. He timed how long it took them to pass between two of the peaks. When he did the calculations later, he discovered that they must have been travelling at a rate of 2,735km (1,700 miles) an hour, more than twice the speed of sound. The first supersonic flight by an aircraft was still four months in the future and the first plane to fly twice the speed of sound didn't do so until 1953.

When Arnold described his sighting to other pilots some thought they might have been some sort of secret, experimental guided missile, while others were reminded of the foo fighters of the recent war. Arnold took his report to the nearest FBI office but, finding it closed, went on to the local newspaper. He told a reporter there that the objects "flew like a saucer would if you skipped it across the water". That description was a vivid image (even though Arnold never said that the objects themselves were saucer-shaped; he described them as looking like crescents) and when the news services picked up the story, they reported on the "flying disks", "flying platters" and, for all posterity, "flying saucers".

In the 70 years since Arnold's apparently detailed experience, the flying saucer mythology has grown to epic and unbelievably complex

ARE WE BEING VISITED?

While there is almost certainly life – and possibly intelligent life – elsewhere in the universe, the question is: are we being visited by its representatives?

The answer is, sadly, probably not. Both logic and physics conspire to make alien visitors – at least the sort that we read about in the tabloids and most pro-UFO books – incredibly unlikely. There are really only two possibilities. Life is either found throughout the universe or it is a rare phenomenon. If the universe is teeming with life, then there can really be nothing especially unique about our own planet, certainly not enough to warrant the amount of attention we seem to be receiving. After all, our universe is nearly 14 billion years old. If life exists throughout it, it must exist in every stage of development, from primitive single-celled organisms to beings who have evolved thousands and even millions of years beyond us. Why would Earth especially be singled out?

This bears upon the common science fiction theme of extraterrestrial invasion. What in the world would Earth have that could not be closer to home? We know that the elements can be found throughout the universe – everything from iron to water and even complex hydrocarbons. Why travel hundreds of light years to obtain what could be found in one's own backyard? Interstellar travel – even for the most technologically advanced races – would be an immensely expensive proposition. The investment in time, energy and resources that it would take to invade our world – for any reason – would

hardly be worth making. On the other hand, life may be vanishingly rare. In that case, the discovery of a planet teeming with life – like our own – would be of paramount importance, perhaps even the most important thing to happen in the history of a civilization. It certainly would not be an event that would be treated casually. It would be as important to them as it would be to us. For the same reason extraterrestrial invasions are unlikely, alien visitors would hardly make such an immense investment and then fool around, showing off fancy aerobatics and playing peekaboo.

The bottom line would seem to be that there is no reason to suppose that Earth is the Grand Central Station of the universe, as it would seem many UFO proponents would have us believe.

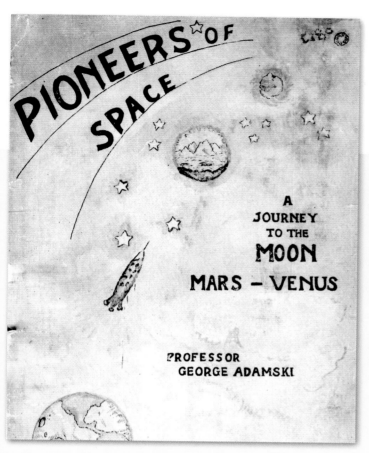

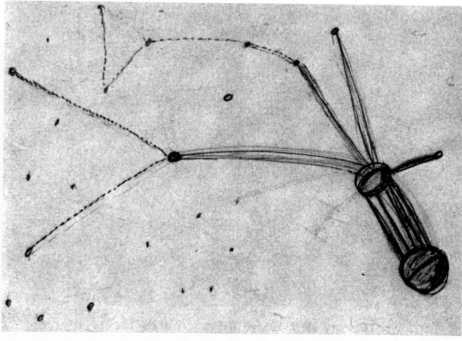

OPPOSITE: A Texas student, Mary Ellen Anderson, gazes skywards at one of the "Flying Saucer Observation Posts" established in the postwar United States.

ABOVE LEFT: Much of George Adamski's "factual" encounter with the inhabitants of a UFO was taken almost verbatim from a science fiction novel he had self-published several years earlier, in 1949.

ABOVE RIGHT: This map drawn in 1961 by UFO abductee Betty Hill purports to show interstellar trading routes. Astronomers have shown that it can be aligned with almost any region of stars chosen at random.

proportions, so much so that it is not possible to do more than provide the briefest synopsis in the space of a single chapter. It is just as well, though, since the subject of UFOs and the subject of extraterrestrials are largely parallel stories. Our interest here is less in the machines than it is in the creatures that operate them. And while there have been tens of thousands who believe they have seen those machines, there is only a handful who believe they have met the occupants face to face.

The Contactees...

The owner of a California hamburger stand was watching the Orionid meteor shower on an early October evening in 1946. He was astonished to see "a large black object, similar in shape to a gigantic dirigible" pass above nearby Palomar Mountain – home at that time to the world's largest telescope. In August the following year, he claimed to have witnessed nearly 200 saucers flying over the mountain.

"Professor" George Adamski became a celebrity. He had never been slow to pursue what might be a good thing. He had begun by operating a "Tibetan" monastery on the mountain, which, during Prohibition, gave him license to make as much wine as he could – all for "religious purposes". After the repeal of Prohibition, he had drifted into lecturing on astronomy and Eastern philosophy to supplement what he was earning from his roadside hamburger stand. Now, with curiosity about the reported existence of extraterrestrial "flying saucers", he lost no time in jumping on the bandwagon.

Adamski reported having taken photos of the saucers he had seen. He founded an "Advanced Thinkers Club" and lectured widely on his sightings. He wrote and self-published a science fiction novel, *Pioneers of Space* (1949). In it a small group of adventurers build a rocket ship to explore the Moon, Mars and Venus, where they meet the inhabitants of these planets. As utterly dreadful this book is as a work of literature,

it bears a close resemblance to the Professor's later adventures.

Since 1950 Adamski had been talking about how much he wished to meet the occupants of the saucers. While driving through the desert east of Palomar in 1952, Adamski reported having seen "a gigantic cigar-shaped silvery ship". Convinced that the craft had come for him, he chased it down. The large craft flew off, but a small "scout ship" was left behind. Ever-diligent, Adamski took photos of it. Looking down, he was surprised to see a man standing nearby. Adamski knew immediately that he was "in the presence of a man from space – A HUMAN BEING FROM ANOTHER WORLD!"

Using hand gestures alone, Adamski was able to learn that the visitor was from the planet Venus and that the reason for his visit was the concern his people had for the radiation from nuclear testing that was leaking out into space.

The small scout ship was one of several carried aboard the larger "mother ship". The visitors from Venus were not the only ones interested in terrestrial affairs. Saucers were coming to Earth from other planets in our solar system as well as from planets orbiting other stars. Among many other things, the Venusian told Adamski about the religion of his people, how some were even now walking city streets in the midst of humans, that some saucers had crashed and some had been shot down. All within one hour, through hand gestures.

After several more encounters with Venusians and even more photo sessions, Adamski collaborated with British filmmaker and author Desmond Leslie to create the book *Flying Saucers Have Landed* (1953). It became an immediate bestseller. It may or may not have some bearing on the story that only a year later Leslie wrote the screenplay for *Stranger from Venus*, an inept reworking of 1951's *The Day the*

ABOVE: A photograph purporting to reveal one of the human-like Venusian visitors who took George Adamski on a guided tour of the planet Venus, which we now know to be a hellish place of acid rains and furnace-like temperatures – not the most tourist-friendly destination.

ABOVE: Adamski created the myth of the "space brothers" – idealized humans (almost invariably blonde and white-skinned) who arrived in flying saucers and whose only interest was in saving Earth from its eventual, and usually nuclear, doom. Over the past half-century this concept has evolved into what amounts to a new religion.

Earth Stood Still – even sharing one of the latter's stars, Patricia Neal, who must have felt an otherworldly sense of déjà vu. Indeed, the makers of *The Day the Earth Stood Still* must have recognized some remarkably familiar territory in both Adamski's book and Leslie's movie. The later film, probably by no coincidence, parallels most of the descriptions of the Venusians from *Flying Saucers Have Landed* as well as their largely benevolent concern with the affairs of our planet.

Only the most diehard believers accepted Adamski's stories at face value. Others found it all too easy to poke holes in their flimsy fabric. What makes Adamski's reports especially questionable is the fact that the aliens he encountered, their physical appearance, their philosophies and their descriptions of their worlds were nearly identical to what Adamski had written in his novel nearly three years before his first

"encounter". The only thing missing from his novel were flying saucers and alien visitors to Earth.

Adamski has been shown to have been a fraud and his saucer photos revealed as fakes, but by the time that happened the damage had been done. Other "contactees" began reporting their experiences, usually following Adamski's lead when it came to what the "space brothers" looked like, where they came from (Venus was a clear favourite) and their Messiah-like message to the people of Earth.

Thus began what flying saucer historian Curtis Peebles calls the "contactee myth". According to Peebles, the flying saucer contactees generally follow an agenda that bears more than a superficial resemblance to a nascent religion. Certain individuals, Peebles points out, are hand-picked by aliens for personal contact. This can be in the form of a face-to-face meeting or via mental telepathy. Sometimes the contactee

"Until they come to see us from their planet, I wait patiently. I hear them saying: 'Don't call us, we'll call you.'"

Marlene Dietrich (1962)

is taken aboard a flying saucer. The really lucky ones are treated to a guided tour of our solar system. Almost invariably, it is revealed that the aliens come from ideal utopias, free of war, hunger and disease. The "space brothers" and "space sisters" are invariably humanoid, tall, handsome, with flowing locks of hair and virtually without exception Caucasian. Their overwhelming motive is to save humans from themselves. To accomplish this end, they impart a message of peace and brotherhood to the contactee, exhorting him or her to spread this good news around the world.

Meanwhile, the mysterious "Men in Black" are intent on preventing this from happening and will do anything up to and including murder to achieve that.

...and the Abductees

Perhaps a little less fortunate than the contactees are the abductees. Instead of a friendly invitation to take a sightseeing trip to Venus in order to be converted into a kind of messiah for all mankind, the poor abductee is kidnapped and usually treated pretty rudely. There are stories of medical experiments, probes, erased or implanted memories and even forced pregnancies, all in all making the abductee's experiences far less pleasant than those of the contactees.

What sets the stories told by the abductees apart from those of the contactees is, by and large, the aliens themselves. The contactee aliens are almost always human or humanoid, often handsome and physically perfect to a fault. However, the abductee aliens are usually far less appealing creatures. And this is where the infamous "grey alien" enters, coming in via

the first and most famous abduction story. It is worth recounting this event in some detail since it is the prototype of virtually every alien abduction case that followed it.

Sometime after midnight on 19–20 September 1961, Betty and Barney Hill were driving through the White Mountains of New Hampshire when they saw what they first thought was a bright star in the night sky. Betty thought that the light seemed to be following the car and urged her husband to stop so she could look at it with her binoculars. Barney stopped the car and went into a field where he could see the light better. Using the binoculars he saw figures moving about inside a glowing object. Seeing that they had noticed him, Barney raced back to the car and drove away. As they escaped they heard a beeping sound from behind them. This was followed by a sense of drowsiness. When they again heard the beeping they were 56km (35 miles) south of where they had seen the UFO with no recollection of the intervening time.

For weeks after the event, they both suffered from continual anxiety and ill-defined fears. For example, Betty was sure that they had been exposed to some kind of dangerous radiation. After reading Major Donald E Keyhoe's book, *The Flying Saucer Conspiracy* (1955), Betty became convinced that the dreams she had been having of aliens taking her and her husband aboard a spacecraft where medical examinations were made on them were not dreams at all. Instead, she believed they were a dimly remembered reality.

At the urging of a journalist who later documented their case, the Hills finally sought professional help from Boston psychiatrist and neurologist Dr Benjamin Simon. Three years after

OPPOSITE: The UFO mania of the 1950s manifested itself in hundreds of consumer products, including this plastic model kit from 1952, the first-ever kit of a flying saucer. It was prominently featured in the notorious cult movie *Plan 9 From Outer Space* (1959).

PLASTIC MODEL
FLYING
SAUCER
BY PAUL LINDBERG

"I think that it is much more likely that the reports of flying saucers are the results of the known irrational characteristics of terrestrial intelligence than of the unknown rational efforts of extra-terrestrial intelligence."

Richard Feynman (1992)

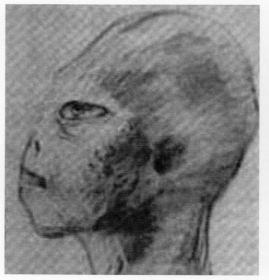

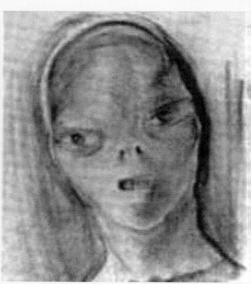

the original event, he began conducting sessions of hypnotic regression during which the couple recounted an experience identical to the dreams they had been having and what little they remembered of the strange encounter. The scenario they told Dr Simon was much more detailed and much more frightening than they had expected. According to Betty, she and Barney had found themselves in the early hours of 20 September on a back road blocked by a dozen aliens. They were a little over 1.5m (5ft) tall, were earless with slit-like mouths, small noses and enormous cat-like eyes that seemed to wrap around the sides of their heads. Their heads were

OPPOSITE TOP: Although there were precedents, the worldwide publicity following the 1961 encounter of Betty and Barney Hill (seen here) with a UFO created the modern "abductee" culture.

OPPOSITE BOTTOM LEFT, CENTRE, RIGHT: Drawings of the aliens that supposedly abducted Betty and Barney Hill popularized the conception of what is now known as the "grey alien". It may not be coincidental that shortly before the Hills described the appearance of their aliens, an episode of the popular television show *Outer Limits* featured an almost identical extraterrestrial. The drawing on the left was done by Barney Hill while under hypnosis. The other two were drawn by hypnotist David Baker from Hill's descriptions.

RIGHT: This photograph that Adamski allegedly snapped in 1952 of one of his UFOs was widely published around the world. The UFO was later shown to have been made from part of a gasoline lantern.

large, tapering down to a small, pointed chin. These creatures led the Hills to a spacecraft that had landed nearby. There, they were taken to separate rooms where they were stripped and placed on examination tables. A needle was inserted into Betty's navel, while a cup-like instrument was placed over Barney's groin. A ring of warts later marked this area. In answer to Betty's questions, one of the aliens showed her a star chart and she saw a book filled with strange writing.

Betty was returned to the car where she saw Barney waiting in a kind of daze. Barney's story was almost identical to his wife's in every detail. But not much can be made of this. They'd had years to discuss their dreams before their sessions with Dr Simon and Betty had made a hobby of reading every book she could find about UFOs.

Betty was careful to point out that there was evidence for their experience other than the memories of herself and her husband.

For example, she had found a dozen "shiny circles" on the car's trunk after the encounter. Betty thought they might have been radioactive so she ran a compass over them. The compass needle moved erratically when Betty did this test, confirming her fears. She assumed these marks had been made when they had heard the strange beeping sounds behind them. Betty said she had made the test with the compass on the advice of a physicist, but this is certainly untrue. A compass can only detect a magnetic field, not radiation.

Betty produced other bits of "evidence", almost all of which is either coincidental or doubtful, but the most compelling to UFO believers is the star map she had been shown.

Betty recreated the map from memory and showed it to amateur astronomer Marjorie Fish, who then tried to match the dots and lines Betty had drawn with existing stars. She believed she found a match in the double star system of Zeta

"It is my thesis that flying saucers are real, and that they are spaceships from another solar system."

Hermann Oberth (1954)

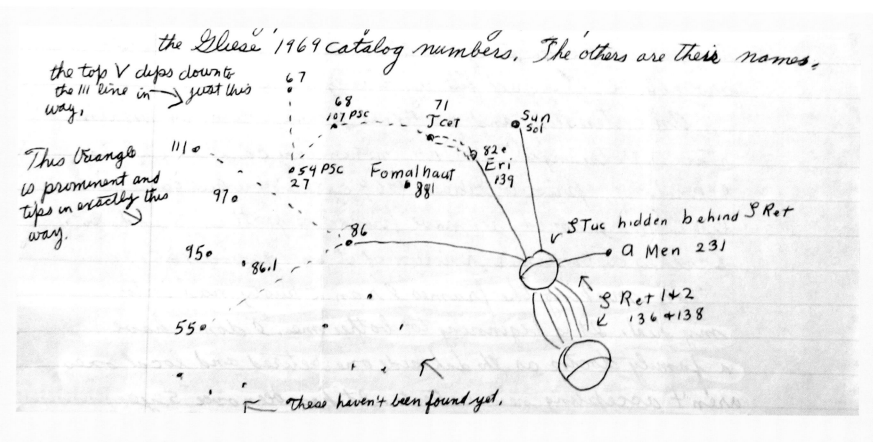

the *Gliese* 1969 catalog numbers. The others are their names.

the top V dips down to the III line in → just this way.

This triangle is prominent and tips in exactly this way.

67
68
107 PSC
71
J Cet
Sun
Sol
111
82
Eri
139
54 PSC
Fomalhaut
27
881
97
86
√ S Tuc hidden behind S Ret
a Men 231
95
86.1
S Ret 1+2
136 +138
55

← these haven't been found yet.

ABOVE: Schoolteacher Marjorie Fish attempted to reconcile the Hill map with known stars. Astronomers have pointed out that not only did Fish cherry-pick the stars she matched, but that the map could be made to fit almost any grouping of stars. It was even shown to match the location of towns and cities in New England!

OPPOSITE: This photograph of a UFO spotted near McMinnville, Oregon in 1950 was among several of the object that were published widely. While UFO sceptics Philip J Klass and Robert Sheaffer believed they had clear evidence of a hoax, UFO researcher Bruce Maccabee disagreed.

Reticuli. But there are many problems with this. Any random collection of vague dots and lines can find a match somewhere on a star chart, especially if there is no indication of scale applied. Also, since the pattern of a constellation depends entirely on the location of the observer, why would the aliens have maps showing their star system as seen from Earth? For that matter, thinking logically, why would they depend on star maps at all? Sceptics have pointed out inconsistencies in Betty's story (Barney largely seems to have gone along with whatever she said). For example, the aliens communicated with Barney telepathically but spoke in English to Betty, and Betty's description of the aliens' appearance changed from telling to telling.

Many of the physical attributes of "grey aliens" can be traced back to the imagery, concepts, events, creatures and identities embedded in mythology and folklore (see Chapter 7). So too can the lurid obsession with sex and human reproduction, which was a key point in the Hills' experience and remains a central part in most alien abduction cases since then.

"Women are yet alive who tell they were taken away when in child-bed to nurse Fairie children," reported Reverend Robert Kirk in *The Secret Commonwealth of Fairies* (1692). While Carl Sagan wrote in *The Demon-Haunted World* (1995) how in tales of ancient gods "most of the central elements of the alien abduction are present, including sexually obsessive non-humans who

LEFT: On the evening of 24 February 1942, shortly after the United States entered World War II, the citizens of Los Angeles were startled by the sight of searchlights in the night sky and the sounds of an aerial barrage – an event caught by a *Los Angeles Times* photographer. Spotters reported having seen mysterious lights, which they believed were enemy aircraft. No one knows for sure what happened that night, but UFO proponents believe that what the military had seen were in fact flying saucers.

live in the sky, walk through walls, communicate telepathically, and perform breeding experiments on the human species." In an honest attempt to understand tales and beliefs like these that have been told and retold for countless centuries, Sagan asks: "Is there any real alternative besides a shared delusion based on common brain wiring and chemistry?"

The notion of "lost time", which is a key feature in both the Hills' encounter and many subsequent cases, can also find its roots in folklore and mythology. In tales of fairy kidnappings, it is common for the victim, who thought only a few hours or days had gone past, to discover that they had been missing for a year or more.

Human encounters with fairies have traditionally also had physical and mental side effects, not all of which are pleasant. There have been reports in fairy lore of hallucinations, insanity, the loss of an eye as well as skin problems inflicted on victims. One thinks immediately of Barney's ring of warts, which is also reminiscent of the "fairy rings" often found in woods and pastures.

Science fiction, folklore and the then-emerging UFO mythology, combined with fears generated

by the Cold War, the atomic bomb and the civil and racial unrest of the time – to say nothing of the Hills', and especially Betty's, psychological condition – were all exacerbated by suggestive hypnosis sessions that appear to have implanted as many "memories" as they uncovered. All of this would seem sufficient to account for this seminal alien "encounter".

While all varying in their details, most accounts of alien abductions in the half century following the Hill adventure follow a familiar pattern – there have been many thousands of such reports from around the world. Common to most are descriptions of being taken aboard a spacecraft, of being stripped of clothing, of operating rooms and invasive medical procedures. There are hours of missing time, gaps in memory, mysterious injuries, implanted devices, disorientation and medical/ reproductive issues. There are even victims who report having been abducted multiple times. Sexual activity is a common theme. Examinations and probes are almost invariably sexual. Sometimes eggs or sperm are harvested or the victims are forced to have intercourse with other prisoners or even their alien captors.

Many UFO believers find these stories compelling, but what should we make of them? The victims all seem to be sincere and usually deeply moved by their experiences. But this simply means that *they* believe in the truth of their stories, not that the stories themselves have any reality. This is why so many abductees pass lie detector tests: these only work when the test subject *knows* he or she is lying.

Another problem is that in all of the thousands upon thousands of reported cases of alien abductions, there is very little hard evidence to work with. No one has managed to escape with even the smallest souvenir of their adventures – even when a victim has been allowed to wander freely and unsupervised around the spacecraft. In fact, a point is often made of how they are specifically forbidden or prevented from taking away any physical evidence of their experience. What about the physical evidence of their own bodies: scars, incisions, bruises, warts and other markings? These are not evidence of an extraterrestrial encounter. We also have to take into account the fact that much of the "science" imparted by the alien visitors – astronomical, biological, physical, etc. – is simply nonsense. For example, when humans have been taken on visits to planets of our own solar system, they describe having seen worlds completely at variance with what we know conditions there to be, such as a balmy, Eden-like Venus or a forested moon. If these stories do not ring true scientifically, what is their origin and how have they become embedded in our mythology?

TAKEN TO VENUS

"T. Lobsang Rampa" (the pen name of British author and mystic Cyril Hoskin) was supposedly taken by a spaceship to Venus, which he reported in his aptly titled book, *My Visit to Venus* (1957). There he witnessed "fairy cities reaching up into the sky, immense structures, ethereal, almost unbelievable in the delicate tracing of their buildings. Tall spires and bulbous cupolas, and from tower to tower stretched bridges like spider's webs, and like spider's webs they gleamed with living colours, reds and blues, mauves and purples, and gold, and yet what a curious thought, there was no sunlight. This whole world was covered in cloud. I looked about me as we flashed over city after city, and it seemed to me that the whole atmosphere was luminous, everything in the sky gave light, there was no shadow, but also there was no central point of light. It seemed as if the whole cloud structure radiated light evenly, unobtrusively, a light of such a quality as I had never believed existed. It was pure and clean…. At last we left the cities and came to a beautiful sparkling sea, a sea of purest blue. There were a few little craft upon the water…"

Since the surface temperature of Venus is *c*370°C (700°F) and it rains sulphuric acid, most modern abductees avoid this problem by having their extraterrestrial experiences take place on planets outside our solar system.

A GALLERY OF FLYING SAUCERS

In the three-quarters of a century since Kenneth Arnold ignited the flying saucer craze, there have been hundreds of photos taken purporting to show alien spacecraft. Most have proven to be either hoaxes, misidentified aircraft or natural phenomena. A rare few have defied analysis. One thing is for certain – such photographs have a powerful draw on our imagination and fear.

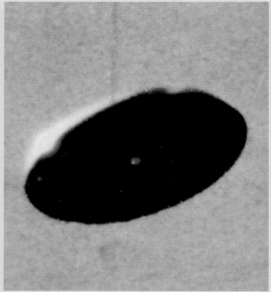

ABOVE: A striking colour photograph of a UFO seen over Painted Lake, Wisconsin.

LEFT: This famous photograph of UFOs hovering over a town in Sicily has been published widely ever since it first appeared in the early 1960s in spite of being discredited as an obvious hoax.

TOP LEFT: In 1963 the Los Angeles-based Amalgamated Flying Saucer Clubs of America released this photograph reportedly taken by one of its members. Almost too good to be true in its clarity, it appears to have been a well-crafted hoax.

BOTTOM LEFT: This classic UFO was seen in the skies over Brazil in the early 1960s.

TOP RIGHT: Several photographs of this UFO were taken by highway inspector Rex Heflin near Santa Ana, California in 1965. Unfortunately, modern photo analysis reveals the saucer to be a fake.

BOTTOM RIGHT: A New Mexico State University geology student claimed to have captured this image of a spherical UFO while photographing land formations in 1967.

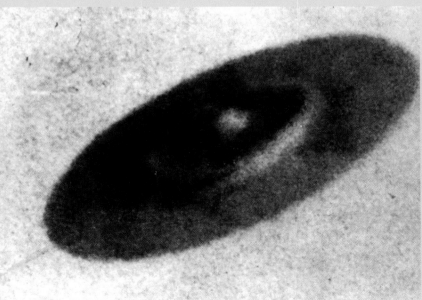

07

FOLKLORE AND EARLY SCIENCE FICTION ALIENS

OPPOSITE: Artist Frank Nankivell created this political cartoon for *Puck* magazine in 1901, featuring a spectacularly effulgent Martian ambassador visiting Washington, DC. Even though it was meant to be humorous, his depiction of the Martian as being more or less human is in line with the expectations of the time.

Aliens living on other worlds were once thought to be not particularly different from Earth's human beings. As we have discovered, the earliest stories involving either trips to other worlds or visits to Earth by alien beings almost invariably described them as being distinctly humanoid and often "superhuman".

It is clear that the first authors to write about aliens from other worlds were, on the whole, much less interested in any kind of scientific accuracy – or even scientific speculation – than they were in using their fictional aliens as a mirror of human life and society, and most stories about travel to other planets were meant to be satirical. One of the earliest examples is "Micromégas", a short story published in 1752 by the French philosopher and satirist Voltaire. The title character is a visitor to Earth from a planet circling the star Sirius. As his home world is 21.6 million times larger than Earth, Micromégas is an impressive 37km (23 miles) tall. On his way to Earth, he stops at Saturn to take on a companion, a relative midget only a mile in height. Although the aliens' visit to Earth was meant – as *Gulliver's Travels* was later – to allow Voltaire to look at earthly governments, society and morals with a satirical eye, "Micromégas" does make one prescient

observation: that alien life may be so different from our own that we may not even recognize it when we see it. For example, Micromégas and his companion are so huge that most terrestrial life – human beings included – is almost too small to be seen by them. They come to the conclusion that Earth is in fact devoid of life.

While most authors wrote of more or less humanoid life living on the Moon and planets, there were a handful of exceptions. In Captain Samuel Brunt's *A Voyage to Cacklogallinia* (1727), for example, the hero is taken to the Moon by the inhabitants of Cacklogallinia: a race of giant, intelligent chickens.

What Do They Look Like?

Most writers at the end of the 19th and well into the 20th century held the idea that aliens would be at least humanoid. Even after H G Wells terrified readers with his description of cold-blooded Martian octopuses, writers like George Griffith could describe a journey through our solar system in *A Honeymoon in Space* (1901) in which Venusians, Martians and Ganymedans all vary but little from the standard human mould. Likewise, Percy Greg's *Across the Zodiac* (1880), Robert Cromie's *A Plunge into Space* (1890),

"A circle of fire coming in the sky, noiseless, one rod long with its body and one rod wide. After some days these things became more numerous, shining more than the brightness of the sun."

From a series of Egyptian hieroglyphs on a papyrus dated to the reign of Thutmose III (1504–1450 BCE)

ABOVE: A fossilized Martian is discovered in Henri de Parville's 1865 novel, *An Inhabitant of Mars*. The appearance of the Martian – large round head, large almond-shaped eyes, small features and fragile-looking body – makes it one of the earliest prototypes of today's "grey alien".

John Munro's *A Trip to Venus* (1897) and Fenton Ash's *A Trip to Mars* (1909) all imagined the planets to be inhabited by thoroughly human creatures, though sometimes idealized to a point where they are almost indistinguishable from angels.

Meanwhile, other authors laid the groundwork for the modern, typical incarnation of aliens: the famous "greys" with their small, typically grey-coloured bodies, slight limbs, bulging heads and enormous eyes. One of the earliest appearances of this now-familiar alien archetype was in the 1865 novel *Un Habitant de la Planète Mars* (*An Inhabitant of the Planet Mars*), written by young French science editor and journalist, Henri de Parville (the pen name of François Henri Peudefer). In this novel the remains of a Martian are fortuitously discovered embedded in an ancient meteorite. The mummified creature bears a striking resemblance to aliens many claim to have seen today.

The insect-like inhabitants of the Moon described in H G Wells' *The First Men in the Moon* (1901) bear many hallmarks of today's aliens. Wells' description of a typical Selenite has some familiar details:

…a trivial being, a mere ant, scarcely five feet high. He was wearing garments of some leathery substance, so that no portion of his actual body appeared, but of this, of course, we were entirely ignorant. He presented himself, therefore, as a compact, bristling creature, having much of the quality of a complicated insect, with whip-like tentacles and a clanging arm projecting from his shining cylindrical body case. The form of his head was hidden by his enormous many-spiked helmet – we discovered afterwards that he used the spikes for prodding refractory mooncalves – and a pair of goggles of darkened glass, set very much at the side, gave a bird-like quality to the metallic apparatus that covered his face. His arms did not project beyond his body case, and he carried himself upon short legs that, wrapped though they were in warm coverings, seemed to our terrestrial eyes inordinately flimsy. They had very short thighs, very long shanks, and little feet.

A little later, Wells' hero sees a Selenite face for the first time, and provides a description with some familiar characteristics, and an unsettling abbreviation of features:

There was no nose, and the thing had dull bulging eyes at the side – in the silhouette I had supposed they were ears. There were no ears…

RIGHT: A modern vision of the intelligent and mysteriously powerful extraterrestrial alien is rendered here beautifully, and hauntingly, by artist Stephen Hickman.

"The phenomenon of UFOs does exist, and it must be treated seriously."

Mikhail Gorbachev (1990)

But it is the Grand Lunar, the vast intelligence that rules the Moon, that is of even more interest. This being is described with a sense of ineffable mystery and with the sense of being transfixed by a penetrating gaze:

He seemed a small, self-luminous cloud at first, brooding on his sombre throne; his brain case must have measured many yards in diameter. . . . At first as I peered into the radiating glow this quintessential brain looked very much like an opaque, featureless bladder with dim, undulating ghosts of convolutions writhing visibly within. Then beneath its enormity and just above the edge of the throne one saw with a start minute elfin eyes peering out of the glow.

No face, but eyes, as if they peered through holes. At first I could see no more than these two staring little eyes, and then below I distinguished the little dwarfed body and its insect-jointed limbs shrivelled and white. The eyes stared down at me with a strange intensity, and the lower part of the swollen globe was wrinkled. Ineffectual-looking little hand-tentacles steadied this shape on the throne…

The Martians discovered by the explorer heroes of Arnould Galopin's *Dr Omega* (1906) included a dominant species of dome-headed, hairless gnomes not much larger than a small child, while the beautifully charming illustrations by W R Leigh that accompanied H G Wells' "The Things

OPPOSITE CLOCKWISE FROM TOP LEFT: The *Superworld* comic lasted for only three issues published in 1940 and featured artwork by Frank R Paul, the first illustrator to specialize in science fiction and a veteran of science fiction pulp magazines since their first appearance in the 1920s. *Operation Peril* appeared in 1951 and reflected all of the tropes of the burgeoning flying saucer craze, right down to hordes of bug-eyed aliens. Artist Basil Wolverton made a lifelong specialty of creating bizarre creatures, and this hungry blob, who featured in *Amazing Mystery Funnies*, was not the strangest of them. The eponymous *Robotmen of the Lost Planet* (1952) resembled more than anything else the then-popular rubber "stress relief" dolls, the eyes and ears of which popped out when squeezed.

FAR LEFT: One of H G Wells' astronauts finds himself the prisoner of insect-like Selenites inside *First Men in the Moon* (1901). In this illustration by Claude Shepperson from the first edition, we can see inside their bulging heads and spindly bodies – some of the characteristics of today's alien depictions.

LEFT: Another early fictional character finds himself imposed upon, this time by Martians. The eponymous Doctor Omega discovers a race of diminutive creatures that are yet another prototype for today's conception of extraterrestrials: small beings with child-like bodies, large eyes and hairless heads. This illustration is by Rapeno from a 1949 edition of the book.

COMIC BOOK ALIENS

Comic books, with their long history of science fiction themes, have featured aliens for three-quarters of a century. For the most part they followed the lead of the pulp magazines, with aliens that were only distinguishable from humans by being coloured green or blue and sporting antennae growing from their foreheads and an occasional extra set of limbs. Also like the pulps, the more villainous an alien was, the more horrible it looked. On the other hand, the most benevolent aliens – ranging from Superman to Lars of Mars – were not only invariably human, they were as good-looking as their artists could draw them. It was exceedingly rare to see a genuinely imaginative extraterrestrial, but the kids never seemed to care.

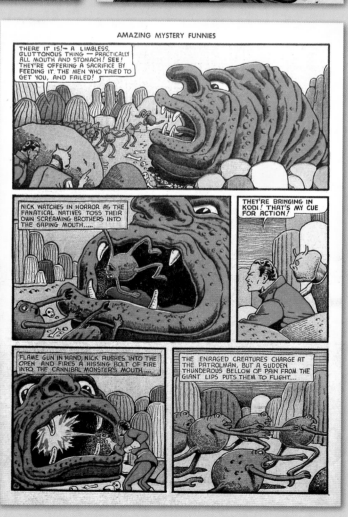

"The cumulative evidence for the existence of UFOs is quite overwhelming and I accept the fact of their existence."

Air Chief Marshal Lord Dowding, Royal Air Force (1954)

(*The Unknown Danger*, 1933) by Gabriel Linde. A race of aliens is described as "short, shorter than the average Japanese, and their heads were big and bald, with strong, square foreheads, and very small noses and mouths, and weak chins. What was most extraordinary about them were the eyes – large, dark, gleaming, with a sharp gaze. They wore clothes made of soft grey fabric, and their limbs seemed to be similar to those of humans."

Our modern expectation of the physical appearance of extraterrestrials can in part trace its origins to stories like these. Mythology and folklore can also claim its share of inspiration as well. Virtually every culture on Earth has some version of the brownie, goblin or fairy, all of which have one thing in common: they are human-like creatures of miniature or even miniscule size. In *The Secret Commonwealth of Fairies*, Robert Kirk describes diminutive beings with "light, changeable bodies . . . somewhat of the nature of a condensed cloud, and best seen at twilight." (Ironically, Kirk died in the same year his book was published after walking over a "fairy mound".) In fact, many UFOlogists have seriously maintained that "grey aliens" and fairies are in fact one and the same creature. The BBC television production of *Quatermass and the Pit* (1958–59), later filmed as *Ten Million Years to Earth* (1967), went so far as to suggest that brownies and hobgoblins were actually only a kind of embedded racial memory of humankind's origins as a Martian slave race under the domination of a master race of intelligent, psychic insects.

Other aspects of folklore may have influenced the modern UFO phenomenon. For example, many details of Betty and Barney Hill's encounter and

LEFT: An illustration from Garrett Serviss' remarkable *Edison's Conquest of Mars* (1898) depicts a large-headed, bald, bulging-eyed and spindly limbed Martian. This portrayal certainly contributed to today's standard model of alien imagery.

OPPOSITE: *My Favorite Martian* was a popular television series in the 1960s. Ray Walston (right) played "Uncle Martin", a distinctly humanoid Martian stranded on Earth when his spaceship crash lands here. Here he is instructing his nephew, Andromeda (Wayne Stamm), in proper Martian behaviour.

That Live on Mars" (1908) depicted feathery, winged Martians with large domed heads, vast, saucer-like eyes and spindly limbs. Garrett Serviss added his contribution to the growing visualization of the "typical" extraterrestrial with the Martians depicted in his *Edison's Conquest of Mars* (1898). These were described as creatures with bulbous, hairless heads, bulging eyes and scrawny bodies. They only departed from today's "little green men" in being 4.5m (15ft) tall.

Especially relevant because it is so much closer to our time is the novel *Den Okända Faran*

"Life is not a miracle. It is a natural phenomenon, and can be expected to appear whenever there is a planet whose conditions duplicate those of the Earth."

Dr Harold C Urey (1952)

"There has been no evidence indicating the sightings categorized as 'unidentified' as extraterrestrial vehicles."

US Air Force official statement (1969)

abduction parallel stories told by people who were involuntary visitors to the fairy world. In those tales the kidnapped victims were also forbidden to return home with souvenirs. The 15th-century *Le Grant Kalendrier des Bergiers* (*The Great Calendar of Bergiers*; 1496) contains graphic depictions of humans who have been captured by demons and Carl Sagan's *The Demon-Haunted World* pointed out that stories of alien encounters share many striking similarities with medieval accounts of demonic harassment.

There has also been a kind of tradition in which different attempts to visualize the human being of the distant future bear marked similarities to alien depictions. In 1891 Kenneth Folingsby published *Meda: A Tale of the Future*, in which human beings are described as having evolved into "tiny gray men with heads shaped like hot air balloons". This may have inspired H G Wells' description of human evolution in his 1893 essay "The Man of the Year Million":

The man of the year million will not be bothered with servants handing him things on plates which he will chew, and swallow and digest. He will bathe in amber liquid which will be pure food, no waste matter assimilated through the pores of the skin. The mouth will shrink to a rosebud thing; the teeth will disappear; the nose will disappear – it is not nearly as big now as it was in savage days – the ears will go away. They are already folded up from what they were, and only a little tip fast vanishing remains to show that ages ago they were long-pointed things which bent forward and backward to catch the sound of approaching enemies. But the brains grow, for they are exponents of the brain, and the great soulful eyes.

In the early 1930s astronomer-artist Lucien Rudaux published an article in which he attempted to depict what he thought the human of the future might look like. This turned out to be a small creature with slender limbs, a huge, bulging, hairless head, large eyes and small features.

In 1939 the daredevil palaeontologist (and probable model for Indiana Jones) Roy Chapman Andrews published a magazine article called "What We'll Look Like in the Year 500,000" in which he reported on some ideas suggested by anthropologist Dr Harry Shapiro, who "thinks our hypothetical future man will have not only a larger brain case but a rounder one as well . . ." Andrews continued: "Doctor Shapiro also thinks that our brows will be smoother with less prominent eye ridges than are common today . . ." Shapiro was furthermore "convinced that decay, wrongly erupted teeth and crowding . . . foreshadow a smaller face, jaws and teeth and the loss of our 'wisdom teeth' and lateral incisors."

LEFT: H G Wells envisioned the human being of the future as being reduced to an enormous head, with only rudimentary features, body and limbs. This was but one example of the equivalency being made between creatures of superior intellect and the development and atrophy of the brain and body respectively.

OPPOSITE, LEFT: French artist-scientist Lucian Rudaux reiterated the idea of the brain dominating the body in the early 1930s when he published his conception of the human being of the future.

OPPOSITE, RIGHT: In 1939 scientists Roy Chapman Andrews and Harry Shapiro published their concept of mankind's future evolution (bottom). The result was not unlike that of any number of Hollywood aliens.

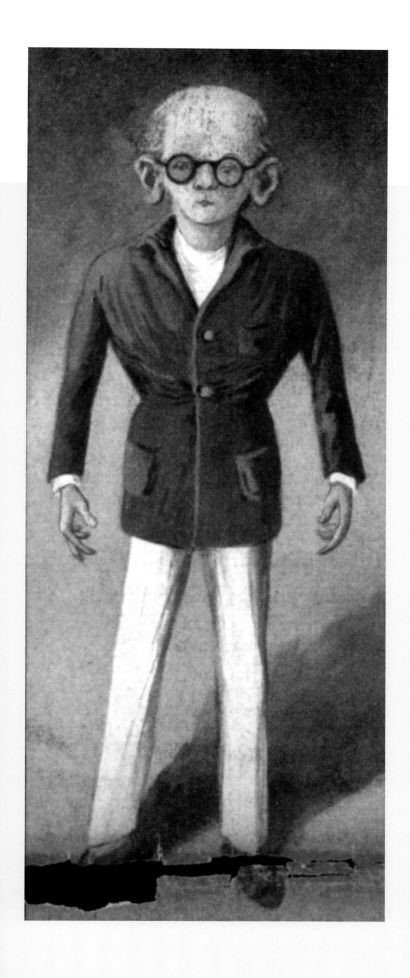

THE FIRST ALIEN CHARACTER IN COMICS

Mr Skygack, a Martian being who came to Earth via a meteorite to study "the ways of the Earth beings", made his first appearance in Chicago's *The Day Book* in 1912, although he had already been featured in other publications as early as 1907. *The Day Book*'s staff cartoonist A D Condo created Mr Skygack in part because it gave him a chance to comment on and satirize the social norms of the day, which was one of the goals of socialist-leaning *The Day Book*. In many of these cartoons we get to read over the shoulder of the Martian as he records his naïve impressions of Earth life. In the words of the cartoon's subtitle, "He Visits the Earth as a Special Correspondent and Makes Wireless Observations in His Notebook".

Mr Skygack was the inspiration for one of the first attempts to create a science fiction alien costume. In 1908 William Fell of Cincinnati attended a masquerade held at a skating rink dressed as Mr Skygack. Later, in 1910, a young woman in Tacoma, Washington created a Skygack costume to wear to a masquerade ball (where she won first prize).

OPPOSITE: Science fiction author E E Smith created a large number of sympathetic aliens in the novels he wrote in the 1930s, '40s and '50s. However, while they might be wildly non-human in appearance, they rarely thought and acted like true extraterrestrials, with alien ideas and alien motives. Rather, they seemed more like humans wearing bizarre costumes.

Future humans would also have no hair, either, whether male or female. "It is true", Andrews adds, "that baldness is much more frequent among highly civilized races of mankind than among the primitive peoples of nature."

One can find countless examples prior to the 1950s where human beings of the future have evolved into creatures who have small, often weak, bodies with slender limbs, huge, usually hairless, heads, large eyes and small features. Perhaps we have become accustomed to associating such beings, as well as aliens in general, who we believe are thousands of years advanced mentally, physically and socially with an ingrained picture of what we think future humans will look like.

Alien Aliens

With the possible exception of Wells' malevolent Martians, no matter how strange an alien was depicted as being it always *thought* like a human. Hollywood, with very rare exceptions, has stuck to this idea tenaciously to this day. Even some of the best modern writers had a hard time conceiving of or expressing the thoughts of a genuinely alien mind. For example, E E "Doc" Smith invented some truly bizarre extraterrestrials for his classic "Lensman" series (1948–54), yet every one of them thought and spoke like a human. The unusual "hero" of Hal Clement's *Mission of Gravity* (1953) is Barlennan, a kind of sentient methane-breathing caterpillar living on an extreme high-gravity planet. The story is related in the

"I think it would be a very rash presumption to think that nowhere else in the cosmos has nature repeated the strange experiment which she has performed on Earth."

Harlow Shapley (1928)

"The Martian wasn't a bird, really. It wasn't
even bird-like, except just at first glance."

Stanley Weinbaum (1934)

first person and had the reader no physical
description of the creature, they would have little
reason to suspect that the story wasn't being
told by a human.

Although Wells had laid the groundwork for a
fundamental change in how alien life might be
realistically perceived, it was a young American
author named Stanley Weinbaum whose short
story "A Martian Odyssey" (1934) contained the
first genuinely *alien* alien. As many authors had
done before in describing extraterrestrial
creatures, Weinbaum made the intelligent,
ostrich-like Martian, Tweel, a physical product of
its environment. But he took this concept a step

further than anyone else had: Tweel also *thought*
like an alien. Weinbaum's unique contribution to
science fiction – and our entire concept of alien
life – was that an alien would be the product of
alien evolution on an alien planet with an alien
environment. It would have an alien biology and
alien senses. Why then would such a creature
think like a human being? In fact, Weinbaum raises
the distinct possibility that an alien might be
different enough to make useful communication
with humans impossible.

In short, Weinbaum introduced two entirely new
ideas to science fiction. Previously, aliens were
depicted as being humanoid and (for the most

ABOVE: Carl H Claudy's *The Mystery Men of Mars* (1933)
featured a number of very imaginative aliens, all strikingly
illustrated by A C Valentine. Claudy was one of the earliest
authors to propose the idea of a non-biological race of
intelligent machines.

ABOVE, LEFT & RIGHT: In spite of a predilection toward humanoid aliens, science fiction illustrators have for more than a hundred years worked hard at depicting ever more unusual and imaginative extraterrestrials. Left: Artist R T Crane depicted the "thinking and talking trees" that were described as living on Venus in the book *Men of Other Planets* (1951). Right: In 1939 the British science fiction magazine *Fantasy* featured an invasion of giant caterpillars in this dramatic painting by S R Drigin.

part) more or less like humans in character and intelligence. Alternatively, aliens were depicted as thoroughly repulsive monsters that wanted nothing more than the conquest or – better yet – obliteration of *Homo sapiens*. Instead, Weinbaum suggested that aliens might be at the same time wholly non-human in every conceivable respect, physically and mentally, yet be neither malevolent or benevolent but instead so *different* as to make their thoughts and motives incomprehensible to humans.

Although "A Martian Odyssey" and other stories Weinbaum wrote that featured plausible alien life forms were a watershed in science

fiction literature, it can still be difficult for an author to imagine a genuinely alien mind and it is still rare to discover an alien in a science fiction novel or short story that has genuinely alien thought processes, motives, emotions and logic. More often than not, we still find fictional extraterrestrials who are the products of alien environments, but which still have understandable, even human, thought processes. A similar problem is faced by authors who hope to write about superhumans. How does one imagine *how* a human being thinks who is a thousand times more intelligent than you are? Probably only one author ever made a successful attempt at

creating a believable superman. This was Olaf Stapledon's creation, John Wainwright, in *Odd John* (1935). It is interesting to note that the depiction of John on the early covers of the book bears a striking resemblance to many of the "Martians" and "Venusians" reported to have visited Earth during the 1950s. Or, for that matter, the Metalunans in the 1955 film, *This Island Earth*.

Even if science fiction authors have struggled to render the truly alien mind, they have explored what seems to be every conceivable possible form intelligent life might take. Kendall Foster Crossen introduced intelligent trees in "The Ambassadors from Venus" (1951). Probably the most disturbing alien plant is the semi-sentient "Mother" in Philip José Farmer's 1951 story of the same name. In *Solaris* (1961) Stanisław Lem described a planet covered by what amounts to an intelligent ocean. Following Weinbaum's lead, Lem's human characters are never able fully to understand –

or even make useful contact – with such an incomprehensibly alien being.

Other fictional aliens simply have no physical form at all. They may lurk as an unseen presence – such as the alien manipulators in Arthur C Clarke's *2001: A Space Odyssey*, or they may be vast intelligences existing in the form of fields of energy. The eponymous being in Fred Hoyle's *The Black Cloud* (1957) is an intelligent cloud of plasma that drifts through our solar system and is able to communicate with human beings telepathically.

Certainly, science fiction authors have gone far beyond the humanoid aliens illustrated in the 19th century or Hollywood's early depictions. Only the most rabid UFO believer really expects aliens to resemble humans, let alone anything else that might have evolved on this planet – and as we know, Earth has, over the past 3 or 4 billion years of its life, produced some pretty bizarre life forms all on its own.

ABOVE: We never see the aliens in Stanley Kubrick's classic film, *2001: A Space Odyssey* (1968). We are shown only their influence on the development of Homo sapiens and the device by which they accomplish this: the famous monolith. One of the movie's basic themes was one that had already been explored by numerous science fiction authors and UFO proponents: that aliens have had a direct hand in directing human history.

OPPOSITE: Many science fiction authors have even abandoned the idea of aliens possessing corporeal bodies, instead portraying them as being masses of pure energy or even the embodiment of thought itself. In this illustration, future space explorers are encountering such a being, which appears as nothing more than a glowing cloud of energy.

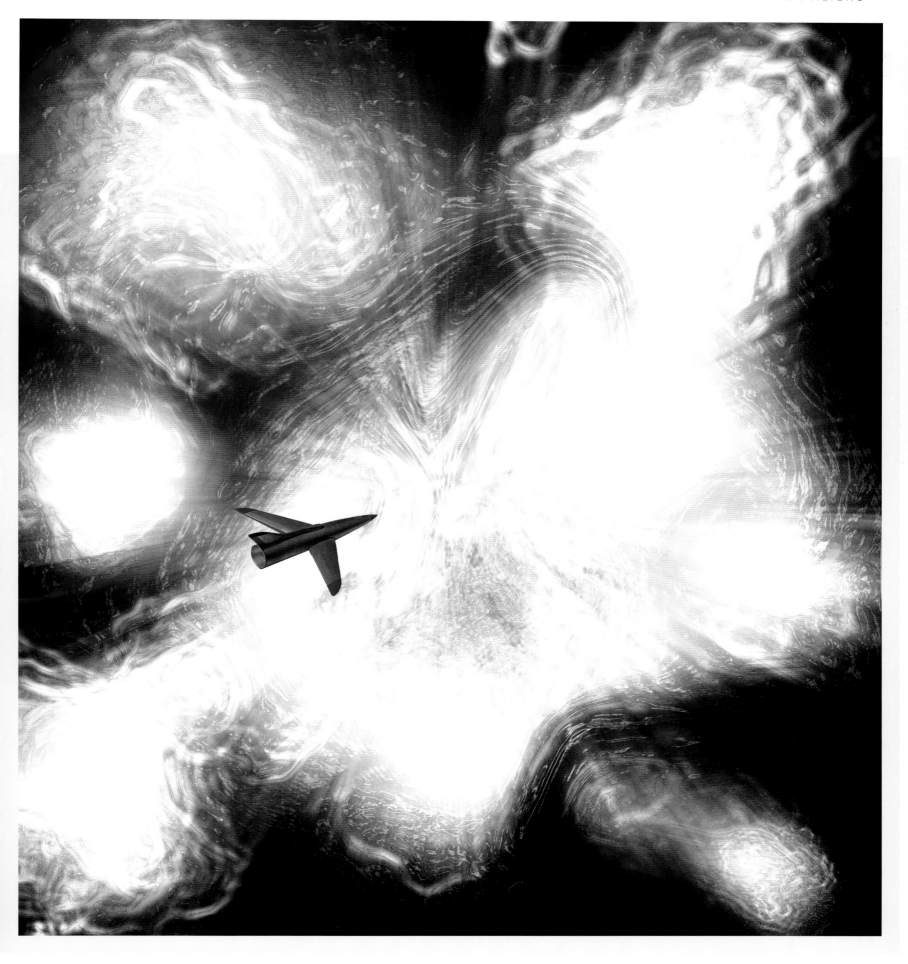

A GALLERY OF PROTO "GREY ALIENS"

The familiar "grey alien", with its bulging head, large eyes, tiny features and small stature, did not spring fully fledged alongside the UFO phenomenon. Instead, it is the product of a long tradition that draws from both folklore and science fiction.

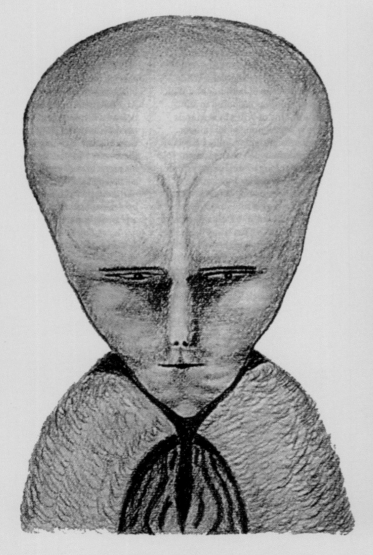

OPPOSITE, BELOW LEFT: The infamous occultist Aleister Crowley created this drawing of a non-terrestrial intelligence named "LAM" in 1916.

OPPOSITE, BELOW RIGHT: This photograph purporting to show a captured alien appeared in a German magazine in the 1950s. It was later revealed to be a hoax.

BELOW LEFT: The bizarre extraterrestrial of *The Man from Planet X* (1951) was clearly an ancestor of today's alien stereotype.

TOP RIGHT: Typical "greys" inhabited the Mushroom Planet in a series of charming children's books by Eleanor Cameron, beginning with *The Wonderful Flight to the Mushroom Planet*, published in 1954.

BOTTOM RIGHT: The idea of the bulbous head and withered body was taken to an extreme in *Invaders from Mars* (1953), in which the Martian mastermind has been reduced to little more than a head with a few ineffective-looking tentacles.

08

ALIENS IN POPULAR CULTURE

OPPOSITE: A scene from Georges Méliès classic science fiction film *A Trip to the Moon* (1902). The aliens were actually played by French acrobats.

A liens have not only been a persistently popular theme in movies for the past century, they also appeared in the first science fiction movie ever made. Georges Méliès' 1902 epic, *A Trip to the Moon*, was a blend of Jules Verne's *From the Earth to the Moon* (1865) and H G Wells' *The First Men in the Moon* (1901). It was from the latter that Georges Méliès took the intelligent insects that ruled the lunar kingdom. The miniscule creatures – played by costumed French acrobats – attack Méliès' explorers, who have to fight them off with their walking sticks and umbrellas.

Méliès and other directors, such as the Spaniard Segundo de Chomón, made movies in which the inhabitants of other planets were depicted. But the burlesque nature of these films – as wonderful as they are – places them in the realm of cartoons, where the extraterrestrials are included more for their grotesque or comic effect than anything else. It was to be several decades before filmmakers made even a half-hearted attempt to depict a realistic alien being.

Aliens Emerge on Film

The first movie to feature an alien from another planet visiting Earth was probably *When the Man in the Moon Seeks a Wife* (1908), the title of which neatly sums up the entire 15-minute-long plot. An entirely human-looking Selenite invents an antigravity gas that allows him to travel to Earth in order to meet an English girl he had spotted through his telescope.

Himmelskibet, a 1918 Danish film released to English-speaking audiences as *A Trip to Mars*, featured some impressive special effects, but the Martians the explorers discover are distinctly human. The year 1920 saw the release of *Algol*, in which the female spirit of the star Algol creates a machine that affects the life of an Earthling. While not a physical alien, Algol is still one of the earliest interstellar visitors in cinema.

"Klaatu barada nikto."

Klaatu, *The Day the Earth Stood Still* (1951)

"I can't think how else to describe it – a horrible, grotesque face looking right into my eyes!"

Enid Elliott, *The Man From Planet X* (1951)

LEFT: The classic Russian film *Aelita* (1929) was set largely on Mars. Although Martian costumes and architecture are as unworldly as one might want, the Martians themselves are solidly human.

OPPOSITE: The ethereal creature named "Ek" who visits the Earth in *One Glorious Day* (1922) may or may not be of extraterrestrial origin, but he certainly anticipates many of the features of the modern alien stereotype, such as the small, delicate body, bald head and enormous eyes.

The incredibly rare 1922 silent movie *One Glorious Day* merits some special attention for its depiction of "Ek". Although the storyline described Ek as the "soul of an unincarnated being", he is recognizable as the precursor of today's infamous "greys" – the large-eyed, small-bodied, hairless beings made famous in Steven Spielberg's *Close Encounters of the Third Kind* (1977). Ek's egg-like head even features a single wick-like antenna. In spite of his semi-spiritual nature, Ek is described as living on the "edge of the Universe", from which he "dives into space and down among the whirling spheres toward Earth".

In 1924 Russian director Yakov Protazanov released *Aelita*, a pro-Soviet epic with parallel stories taking place on Earth and on Mars. Like *Himmelskibet*, there is no attempt to depict a genuinely alien race – though the costumes created for the Martians by Alexandra Exter are famously impressive and very definitely not of this planet. *The Man from M.A.R.S.* (1922) also featured humanoid Martians (with names like Pux Pux and Gin Gin) – but did at least introduce the concept of teleportation as a device for space travel.

Just Imagine (1930), a bizarre science fiction musical, harked back to the burlesques of Georges Méliès: a rocket landing on Mars reveals entirely human Martians – although the twin queens, Loo Loo and Boo Boo (played by the amazing Joyzelle Joyner) – were worth the price of admission. The alien races confronted by

LEFT: *Just Imagine* (1930) was a musical comedy set in the distant year of 1980. It features a rocketship trip to Mars, where the thoroughly human queen of the planet, Loo Loo (and her twin sister, Boo Boo), is played by the flamboyant Joyzelle Joyner.

OPPOSITE: Legendary film production designer William Cameron Menzies directed the low budget but stylish *Invaders from Mars* (1953), which featured the bug-eyed, velour-suited Martians seen here.

Buck Rogers and Flash Gordon in their respective serials were cast in the same mould – and the men from Mars and other worlds that appeared in later serials, such as *Radar Men from the Moon*, et al., followed suit.

There were no genuinely alien extraterrestrials in motion pictures until the creepy little title creature of 1951's *The Man from Planet X*. Even classic films such as *The Day the Earth Stood Still* (1951) and semi-classics such as *Rocketship X-M* (1950) depicted aliens that were entirely human-looking. However, *The Man from Planet X* featured a being that was distinctly otherworldly, and, like Ek nearly three decades earlier, a being that foreshadowed the familiar "greys" of today's UFOlogy. Unfortunately, the mask-faced man

from Planet X remained unique for many years as Hollywood returned to aliens that were for the most part entirely human – or at least more or less human-appearing, such as the man-shaped ambulatory vegetable that terrorized the crew of an arctic base in *The Thing from Another World* (1951). This was probably due as much to a lack of imagination as it was to economics. Human or humanoid aliens are simply less expensive to create.

Nevertheless, there were a number of creditable efforts to depict aliens as being genuinely alien. While most of the Martians in *Invaders from Mars* (1953) were bulky, pop-eyed thugs in velour union suits, they were controlled by a super-intelligence that took the form of a diminutive creature with a bulging head,

ABOVE LEFT & OPPOSITE: *It Came from Outer Space* (1953) not only managed to feature entirely non-human aliens, but depicted them sympathetically. Rather than wanting to conquer Earth, all they want to do is get away as quickly as possible, having crashed on this world accidentally.

ABOVE RIGHT: Created by art director Albert Nozaki, the Martian in *The War of the Worlds* (1953) makes a brief but impressive appearance.

humanoid face and curling tentacles in place of a body. *It Came from Outer Space* (1953) featured shape-changing aliens that could take the form of humans but in their natural state were utterly repellent – and absolutely alien – cyclopean monsters. It is one of the first movies in which a real effort was made to depict aliens as completely non-human, yet sympathetic at the same time: they were monstrous in appearance only. The film was unique in that its aliens were on Earth entirely by accident and wanted nothing more than to be left alone until they could depart. Designer Charles Gemora created a convincing Martian that was seen all too briefly in George Pal's movie version of *The War of the Worlds* (1953). Although on screen for barely a second, its bulging three-lobed eye, pulsing head and spidery limbs are unforgettable.

A nearly forgotten oddity is *Supersonic Saucer* (1956), a British movie for children that not only features a uniquely memorable alien but also a plot closely reminiscent of *E.T. the Extra-Terrestrial* (1982), which also paralleled the story told in J B Priestley's *Snoggle* (1971). Although the speechless, nameless alien is little more than a shapeless white bag-like thing about a foot tall, with a pair of enormous eyes, it manages to convey a real personality and genuine presence in this unpretentious film. An unusual characteristic is the creature's identity as the saucer referred to in the title, as it is able to transform itself at will into a tiny spacecraft.

During the 1950s, and especially during the Red Scare (when it was feared America had been infiltrated by radical left-wingers) conjured up by Senator Joseph McCarthy, many filmmakers used

"I'm for killing that goddamn thing right now."

Parker, *Alien* (1979)

the science fiction themes of alien takeover, mind control and loss of self and identity to comment on the paranoia that was then sweeping the United States. It was a period in which everyone looked at the strangers in town, and even neighbours and relatives, suspiciously; when everyone feared being brainwashed by Communist agents, and of becoming little more than a cog in a faceless political machine in which individuality was a sin punishable by death. Hence Hollywood produced films such as *It Came from Outer Space* (1953), *Invasion of the Body Snatchers* (1956), *The 27th Day* (1957), *I Married a Monster from Outer Space* (1958), *The Brain Eaters* (1958) and countless others in which alien invaders stood in for the dreaded Red Menace. Not too surprisingly, most of these films drew their inspiration directly from the fledgling UFO culture, so rather than being wholly responsible for shaping the modern concept

of an extraterrestrial, the effect was more like self-perpetuating feedback.

The best example of this was the 1956 "documentary" *Unidentified Flying Objects: The True Story of Flying Saucers*. It came out in the same period that saw such sensational books as Donald Keyhoe's *Flying Saucers are Real* (1950) and Frank Scully's *Behind the Flying Saucers* (1950), George Adamski's *Flying Saucers Have Landed* (1953) and *Inside the Space Ships* (1955), Harold Wilkins' *Flying Saucers on the Attack* (1954), as well as the considerably more sober *The Report on Unidentified Flying Objects* (1956) by Captain Edward J Ruppelt, who had been in charge of the United States Air Force's Project Blue Book (the first in a series of official studies into UFOs). There had also been feature articles and even cover stories in such nationally respected magazines as *The Saturday Evening Post*, *Time*, *Popular Science*, *Life* and *Look*. The

HYNEK'S CLASSIFICATION

The making of *Close Encounters of the Third Kind* (1977) recruited long-time UFO researcher and author J Allen Hynek as a consultant (he also appeared briefly in the movie). It was Hynek who had developed a classification system for UFO sightings from which the title of the film was derived. These are often listed as below, but there are many variations:

- **Nocturnal Lights:** A visual sighting of a well-defined light in the night sky, the motion or appearance of which is not readily explainable.

- **Nocturnal Disks:** A visual sighting of a UFO in the form of a structured light seen at night.
- **Daylight Disks:** A visual sighting of a UFO with distinct shape, generally oval or disk-like, seen in the daytime.
- **Radar Cases:** Unidentified blips on radar screens.
- **Radar Visual Cases:** Unidentified blips on radar screens that coincide with visual sightings.
- **Close Encounters of the First Kind:** A UFO seen nearby but without any interaction between the witness or the environment.

- **Close Encounters of the Second Kind:** A UFO sighting that includes direct interaction or physical evidence.
- **Close Encounters of the Third Kind:** A UFO observed with humanoid or humanlike occupants.
- **Close Encounters of the Fourth Kind:** Abductions.
- **Close Encounters of the Fifth Kind:** Communication taking place between a human and an alien.

ABOVE LEFT: The titular creature from 1979's *Alien*, designed by Swiss Surrealist H R Giger, is the only rival to the "grey" in the popular conception of extraterrestrials. The xenomorph has appeared in three direct sequels and in some form in a dozen other films.

ABOVE RIGHT: Among the most iconic of movie aliens is the insectoid "mutant" of *This Island Earth* (1955). The impressive costume proved to be more expensive than expected so it was only finished from the waist up.

film, which mixed fact and fiction, fanned the flames of what was already a public firestorm.

In the nearly seven decades since the 1950s, there have been literally hundreds of movies, large and small, good and bad, that have included aliens in some form or another, many of them are hardly worth mentioning, such as the seemingly endless clones of *Alien*. In many ways, both *Close Encounters* and *Alien* were unfortunate throwbacks to early pulp fiction, and even the fiction of the 19th century, where extra-terrestrials were either monstrous, inhuman horrors or instead godlike, benevolent humanoids. However, there have been a handful of exceptions that stand out not only because of the uniqueness of their vision or the influence they had on later films, but also because of the impact on our

perception of what aliens in general "ought" to look like on the movie screen.

The movie most dependent on the "real" UFO phenomenon was probably *Close Encounters of the Third Kind*. It deserves special mention for having made the infamous "grey alien" the modern archetype. Although the "grey" made an appearance in the 1975 made-for-television *UFO Incident*, which was based on the Betty and Barney Hill story, the critical and popular success of *Close Encounters* embedded the image in the public mind. Before the film's release, movie and television aliens came in a bewildering variety of guises, depending on a director's imagination and a producer's budget. But in the years since 1977, the image of the alien with spidery limbs, featureless grey skin, enormous cat-like eyes and

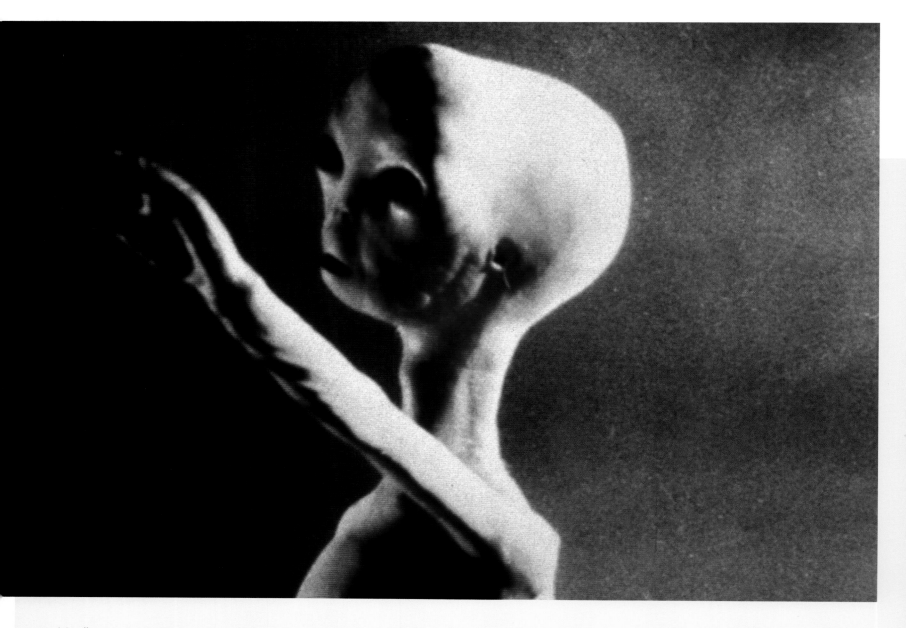

virtually nonexistent nose and mouth has become the norm. In the first decades of human–alien encounters, the aliens looked very much like human beings. Since the time of the Hill encounter, they have almost invariably been depicted as some version wrought from the ubiquitous "grey" (unless they were following the lead of *Alien*). This is especially true for movies involving alien abductions in any way. As, for example, *Fire in the Sky* (1993), *The Fourth Kind* (2009) and the bluntly titled *Alien Abduction* (2014), all of which were based, however loosely, on allegedly real events. Recent films in which the archetypal "grey alien" has appeared include *Extraterrestrial* (2014), in which aliens terrorize teenagers in a remote cabin, *Dark Skies* (2013) and *Paul* (2011), which plays a visit by a "grey alien" to Earth for comedic effect. Popular books like John G Fuller's *The Interrupted Journey* (1966) and movies like *Close Encounters* established a "norm" for the appearance of aliens.

As the public increasingly grew to expect aliens to look like slender grey creatures with tiny features and enormous eyes, Hollywood gave people what it thought they expected to see – an ever-growing, self-feeding spiral that has established in the minds of most people exactly what an alien should look like.

ABOVE: *Alien*'s only rival in terms of public familiarity is 1977's *Close Encounters of the Third Kind*, which cemented forever the physical features of the "grey alien" in the public consciousness.

"It wasn't the right time for us to meet. But there'll be other nights, other stars for us to watch. They'll be back."

John Putnam, *It Came From Outer Space* (1953)

ABOVE LEFT: In 1938, the British science fiction magazine *Tales of Wonder* featured on its cover a civilization of amphibious, frog-like aliens.

ABOVE RIGHT: The tables are turned on the humans in this magazine cover from 1952 as a curious alien wonders at what sort of strange creatures it has found.

A Typology of Aliens

We might be able to divide most science fiction and Hollywood aliens into distinct groups. The first, of course, are aliens who appear to be entirely human, at least in terms of outward appearance. Most of the alien species in *Star Trek* fall into this category, for example. The second type might be intelligent but utterly non-human. The third might be non-intelligent or non-reasoning invaders, such as the eponymous Blob in the 1958 movie or the Ymir from *Twenty Million Miles to Earth* (1957). This category might also include alien diseases such as the one found in *The Andromeda Strain* (1971). A fourth category might be shape-changing aliens, creatures who have no definite form of their own but can take on whatever shape is most convenient for their surroundings. A fifth group might include intelligent parasites who require human (or other) hosts in order to survive. Finally, we might include artificial intelligences and non-organic or incorporeal species. There have been literally hundreds of aliens of virtually every description in movies over the past century. It would be both exhausting and pointless to list every one – or even the most outstanding. But it will be worthwhile to point out a few representative examples of each type.

Aliens Who Appear Human

Certainly the best known of all Hollywood aliens is Klaatu, from the classic *The Day the Earth Stood Still* (1951). Coming from an undisclosed planet, the Christ-like extraterrestrial easily passes for human once he exchanges his spacesuit for a business suit.

A memorable alien who was most definitely humanoid was Nyah, the titular *Devil Girl from Mars* (1954). Inverting the usual story (summed up by the title of 1967's *Mars Needs Women*), it is the men of Mars who are dying this time and Nyah has been assigned to collect a fresh supply of healthy, virile men from Earth. For some reason her choices are reluctant to go.

In *The Man Who Fell to Earth* (1976), David Bowie portrays Thomas Jerome Newton, an alien who has arrived on this planet in search of water to save his home world and the family he left there. He uses his knowledge of advanced technology to create a profitable business. But as he is preparing to leave Earth, he is discovered

ABOVE: Klaatu in *The Day the Earth Stood Still* (1951) was not only an example of Hollywood's continued dependence on humanoid aliens, but was also a predecessor of the "space brothers" soon to dominate the UFO culture: these were god-like aliens who visit Earth only to save us from ourselves.

RIGHT: Despite being yet another humanoid depiction, Nyah, the title character of the movie *Devil Girl from Mars* (1954), as portrayed by Patricia Laffan, was nevertheless a striking portrayal in an effective costume designed by Ronald Cobb.

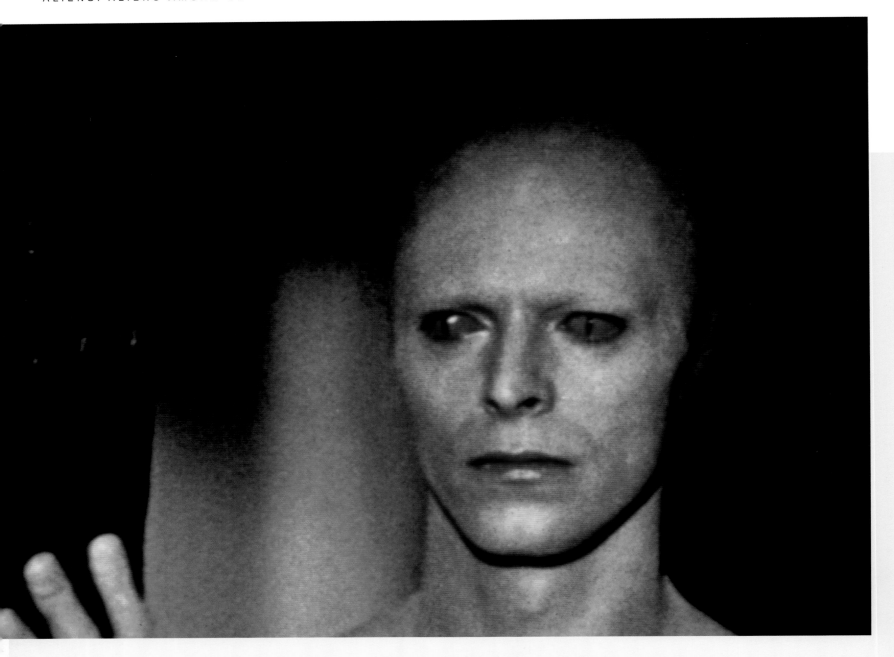

by the government, which threatens the success of his plans and the lives of his wife and children. Newton eventually finds himself lost and abandoned on this planet.

John Carpenter's *Starman* (1984) gave us another sympathetic alien in the nameless visitor who takes on the human form of a woman's late husband in order to escape the pursuing authorities. While this talent may place Starman in the shape-changing alien category, he does spend almost the entirety of the movie in human form. Both the *Star Trek* television series and its

movie franchise have invariably depicted sentient alien species as being humanoid – often strange-looking, but humanoid nevertheless. The archetype, of course, being the iconic Vulcan Mr Spock – even though he *was* half-human.

One of the earliest glimpses of what would eventually become the stereotypical "grey alien" appeared in *Earth vs the Flying Saucers* (1956). A brief shot of one of the aliens' heads revealed a large head set on a small neck, big almond-shaped eyes and a tiny mouth. We are told that the aliens are physically weak and fragile. The

ABOVE: In *The Man Who Fell to Earth*, David Bowie portrayed Thomas Jerome Newton, a humanoid alien who has come to Earth from a world dying of a planet-wide drought.

OPPOSITE: So familiar and popular is the *Star Trek* character Mr Spock that it is sometimes difficult to remember that he is a green-blooded alien (or, rather, half-alien) from the planet Vulcan. Spock's father was a Vulcan while his mother was human, hence his very humanoid features.

"You are a strange species. Not like any other. And you'd be surprised how many there are. Intelligent but savage."

Starman, *Starman* (1984)

ABOVE: In recent years filmmakers have made some effort toward depicting genuinely alien creatures, especially creatures that not only have otherworldly bodies, but also have alien minds and alien motives. Nevertheless, *District 9* (2009) fell short with aliens that were still modelled on the human pattern of one body, one head and four limbs.

OPPOSITE: Firmly rooted in a science fiction tradition going back to the 19th century, the alien of *E.T. the Extra-Terrestrial* (1982) displays all the features that we have come to expect of a space visitor: diminutive stature, small body, spindly limbs, large head and enormous eyes.

face also appeared to be old and wizened, as though belonging to an ancient and perhaps dying race.

As we have seen, *Close Encounters of the Third Kind* (1977) – and to a lesser degree what was essentially its sequel, the immensely popular *E.T. the Extra-Terrestrial* (1982) – firmly implanted the "grey alien" stereotype into motion pictures and public consciousness. Since then, the "grey" in one guise or another has largely become the alien of choice for many filmmakers, especially those who believe they are inserting a degree of "realism" in their movies. This type has been the go-to alien in countless movies and television series, such as *The X-Files* (1993), *Dark Skies* (1996), *Signs* (2002) and *Paul* (2011). One of the

results of this repetition has been to ever more firmly grind the stereotype into the public consciousness as the "right" alien. After all, all of the aliens in movies and on television wouldn't look alike if there wasn't some reason.

The aliens of both *Enemy Mine* (1985) and *District 9* (2009) were obviously the products of evolution in alien environments. For example, the scaly exoskeletons of the "Dracs" in *Enemy Mine* suggest a life form that may have evolved from reptiles on a much more hostile world than Earth, while the "Prawns" in *District 9* obviously had insect ancestors. But both were still humanoid in overall form, with head, torso, two arms and two legs. They were humanoid, too, in intelligence and psychology.

"Welcome to Earth."

Captain Steven Hiller,
Independence Day (1996)

Non-Human Sentient Aliens

Science fiction literature is teeming with utterly non-human aliens. We know that Stanley Weinbaum laid the groundwork for this trend with his "A Martian Odyssey", but he wrote several other stories in which he speculated on not only what kinds of life might evolve in alien environments but what kinds of minds they also may have evolved. In "The Lotus Eaters" (1935), explorers from Earth discover warm-blooded mobile plants with a communal intelligence that might be greater than that of a human. Although capable of developing advanced philosophies and scientific theories, they have no survival instinct and react with indifference when attacked and eaten by the indigenous carnivores. The human explorers are horrified at what they see as an inexplicable behaviour.

Some authors have imagined life based on elements other than carbon. Early descriptions of a silicon-based or crystalline life form can be found in P Schuyler Miller's "The Arrhenius Horror" (1931) and George Wallis' "The Crystal Menace" (1939). John Taine's *The Crystal Horde* (1952) and Isaac Asimov's "The Talking Stone" (1955) both featured intelligent creatures that resembled rocks or crystals.

Some aliens simply have no fixed form. They may be gaseous, resemble flames or even be made of pure energy. The intelligent cloud of hydrogen in Fred Hoyle's *The Black Cloud* (1957) is many times larger than the planet Jupiter, while the eponymous planet in Stanisław Lem's 1961 novel, *Solaris*, is covered from pole to pole by a sentient ocean that proves to be a single biological organism. A similar idea was pursued by Ursula K Le Guin's "Vaster Than Empires and More Slow" (1971) in which she describes a planet covered by lush plants that form a gestalt intelligence. Olaf Stapledon's *The Flames* (1947) and Arthur C Clarke's "Out of the Sun" (1958) both speculated on intelligent beings that resembled living flames.

Any number of stories have focused on aliens that are patterns or fields of pure energy of some sort, such as those in Terry Carr's short story "The Light at the End of the Universe" (1976). Among the earliest of these appeared in Theodore Sturgeon's "Ether Breather" (1939) and its sequel, "Butyl and the Breather" (1940). This category might also include entities that are not just from other dimensions but exist in more than our three dimensions.

ABOVE: Whatever faults *It Conquered the World* (1959) may have had as a movie, at least it could boast of one of the most imaginative aliens to appear on screen up to that time. Created by Paul Blaisdell, the nearly immobile creature was designed to appear as if it had evolved on a planet with heavier gravity than the Earth.

OPPOSITE: In a painting from his book *The Galactic Geographic* (2003), artist Karl Kofoed depicts humans and aliens co-existing amicably.

With only a few notable exceptions, non-humanoid aliens have been a rarity in motion pictures. One reason for this absence may be a general lack of imagination on the part of script writers and creative directors, or the conceit that any intelligent life in space must necessarily look at least remotely human. Or the reason may be even simpler: makeup effects and costumes are expensive, and an advanced alien might break the budget. It's cheaper to use an actor than an actor in makeup, it's cheaper to use an actor in makeup than a full-body costume and it's cheaper to create a full-body costume than a creature that requires puppeteers (like the Alien Queen in *Aliens* of 1986) or advanced CGI (computer-generated imagery). Even one of the most iconic of all movie aliens, the Mutant from *This Island Earth* (1955), wore trousers because

the production didn't have enough money to afford a complete creature. The physical form of the Krell in *Forbidden Planet* (1956) was only suggested by the shape of their wide, rhomboid doorways. They are never seen in the film.

Other films tried harder to understand that alien environments and evolution would produce truly unearthly life forms. The insectoid civilization of *Starship Troopers* (1997) was not only terrifyingly non-human, they even displayed non-human thought processes – a rare accomplishment for Hollywood. An outstanding attempt to depict a wholly non-human species was in James Cameron's *The Abyss* (1989). The semi-transparent, phosphorescent beings in that movie had some of the features of the traditional "grey", but were presented in a way that made them clearly of unearthly origin.

ABOVE: Thanks to computer graphics, Hollywood creature designers are no longer limited to what a human actor can wear when it comes to creating aliens. The insect-inspired "Arachnids" of *Starship Troopers* (1997) are among the most imaginative, convincing and truly alien extra-terrestrials to ever appear on film.

ABOVE: There could scarcely be a less humanoid alien than the titular creature of 1958's *The Blob* – a mindless and ravenous mass of protoplasm with the only goal of eating and growing.

Non-Sentient Aliens

While slightly off-topic, as the focus of this book is really on intelligent or thinking aliens, it would be remiss not at least to mention in passing some of the more interesting non-sentient extraterrestrials. Many of these, and especially those of the 1950s science fiction craze, were created as an excuse to introduce some ravening monster designed to terrorize the local citizens until it could be destroyed by the hero. Although most of these alien creatures were little more than the traditional man-in-a-rubber-suit monsters – such as the Martian creature in *It! The Terror from Beyond Space* (1958) (a film that anticipated *Alien* (1979), which itself was another example of a man-in-a-suit alien, though spectacularly accomplished) – some were imaginative offerings.

The year 1958 saw *The Blob*, a shapeless lump of translucent goo whose only interest was in absorbing humans into its ever-growing mass, and *The Trollenberg Terror* (1958), in which a troop of cyclopean, tentacled horrors from some unknown origin overwhelm a mountaintop observatory. The title creatures of *The Day of the Triffids* (1962), based on the novel by John Wyndham, are enormous, mobile, flesh-eating plants whose seeds arrive on Earth during a meteor shower. A similar theme was brilliantly reprised in Roger Corman's *The Little Shop of Horrors* (1960). Finally, there is the deadly virus in *The Andromeda Strain* (1971), brought to Earth by a returning satellite. The movie reflected a genuine fear that astronauts and spacecraft might bring an alien disease to Earth, one against which terrestrial organisms would have no defence.

ABOVE: Almost as mindless as the Blob, but infinitely more dangerous with their ability to create zombie-like duplicates of human beings, were the seed pods that drifted to Earth in *Invasion of the Body Snatchers* (1956).

For this reason, returning Apollo astronauts – who had been exposed to lunar dust – were thoroughly decontaminated before being allowed in contact with anyone. Likewise, spacecraft designed to land on other planets are rigorously sterilized before being launched into space, lest organisms from Earth contaminate the world being explored. Invasions can work both ways.

Shape-Changing Aliens

Aliens who may appear to be perfectly normal humans, but are not really what they seem, resonate with some of our most primal fears. Clifford Simak's *They Walked Like Men* (1962) featured aliens that resembled nothing more than black bowling balls when in their natural state. By being able to change their shapes at will (they can merge to create objects and creatures larger than themselves) they attempt literally to buy Earth out from underneath the human race by impersonating cash money. An especially creepy scene involves one character's realization that all of the furniture in his apartment might be alive and watching him.

These shape-changers are related in principle to humanoid movie aliens. The difference is that the alien itself is – either by implication or demonstration – wholly inhuman. *Invasion of the Body Snatchers* (1956) is probably the archetype of this in Hollywood – as well as perhaps the best example of an alien invasion movie inspired by the ongoing Red Scare. Based loosely on Jack Finney's novel, the aliens are spore pods that have drifted to Earth from some unknown origin. Each is capable of producing an exact duplicate of a human being, including their memories and

ABOVE: Almost as mindless as the Blob, but infinitely more dangerous with their ability to create zombie-like duplicates of human beings, were the seed pods that drifted to Earth in *Invasion of the Body Snatchers* (1956).

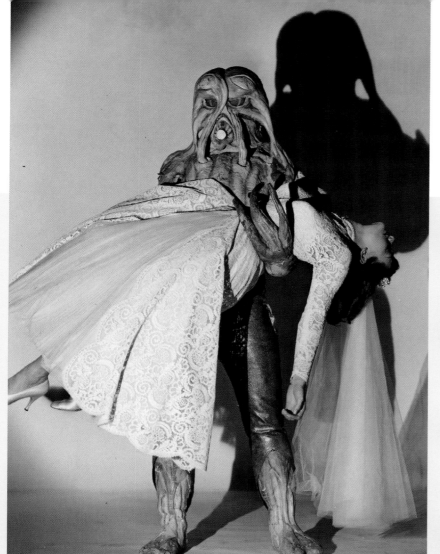

"Humans are driven to search the heavens for company."

Michael Carroll (2016)

ABOVE LEFT: Despite its poor title, director Gene Fowler, Jr's *I Married a Monster from Outer Space* (1958) is in fact a well-made movie featuring an imaginative and, ultimately, sympathetic alien. This was an unusual approach at a time when most movie creatures were bent on world conquest.

ABOVE RIGHT: John W Campbell's 1938 short story, "Who Goes There?", has been officially or unofficially the source of four films, most notably *The Thing from Another World* (1951). A malevolent alien with the ability to take on the appearance of ordinary human beings perfectly suited the Red Scare paranoia of the times.

personalities. However, the replicas are devoid of any emotions. It's never made clear whether the pods themselves have any intelligence or are merely a kind of non-sentient parasite.

I Married a Monster from Outer Space (1958) is a very good little movie suffering under one of the worst titles in film history. Shape-changing aliens from a dying world come to our planet. They hope that interbreeding with our species might help save their race. Director Gene Fowler, Jr did not use this device to avoid showing what the aliens actually looked like but instead revealed their true nature in a series of very effective and even startling scenes.

The Thing from Another World (1951) was based on John Campbell's short story, "Who Goes There?", in which a shape-changing alien finds itself stranded on our world when its spaceship crashes.

It combines its ability to assume the form of any living creature with a frightening intelligence. In the story, the alien is a three-eyed horror. However, in the film the alien is merely an actor (James Arness in one of his first roles) with enough thorns attached to his hands and knuckles to make him appear like the ambulatory vegetable the movie alien was supposed to be. Otherwise, there is not too much to choose in appearance between this alien and Frankenstein's monster.

Under the Skin (2014), based on Michel Faber's novel of the same name, stands out particularly because it is the rare film that tries to see an alien invasion from the point of view of the alien. She is able to take on the outward appearance of a human being, but has no experience in either being human or even how to understand properly her new form.

Parasitical Aliens

Some aliens, while intelligent and capable, are forced to depend on hosts in order to carry out their plans. *The Brain Eaters* (1958) was based on Robert Heinlein's novel *The Puppet Masters* (1951). The story was used without permission and Heinlein sued the filmmakers. Ironically, although the novel has had big-budget movie versions since, the 1958 spin-off may be the most faithful to the book. In both the novel and the movie, slug-like parasites attach themselves to the brainstems of their hosts, thereby controlling their thoughts and actions.

In *The Hidden* (1987) an alien parasite with the ability to possess human bodies goes on a violent crime spree while being pursued by an alien inhabiting a human body. A similar idea was used in *The Brain from Planet Arous* (1957), in which an evil alien intelligence takes over a scientist's brain while the good alien in pursuit takes over the brain of the scientist's dog. Both movies may have been inspired by Hal Clement's 1950 novel, *Needle*, in

which an alien policeman (resembling a 1.8kg (4lb) green jellyfish) with the ability to live in symbiosis within a human being pursues another of his kind hidden somewhere on Earth in another human.

Artificial or Incorporeal Aliens

Finally, we come to the aliens that have no biological form. They may not be even made of matter, but are instead creatures of pure energy or even thought itself.

An early – and outstandingly imaginative – example of the incorporeal alien in motion pictures and television is the villainous Manza, who was terrifying the spaceways in the early live television series *Space Patrol*. The alien, a disembodied voice, could take complete mental control of its victims. In the "The Defeat of Manza" (1954), the being is revealed to be a collection of crystals scattered on the ground in a seemingly random but very specific pattern. The electrical nature of the crystals allowed an intelligence to be created. Unfortunately

ABOVE: The aliens featured in *Invasion of the Saucer Men* (1957) discovered that their enormous brains were no match for American teenagers.

OPPOSITE: The Metal Maniac of Mira Ceti VII, a planet of sentient robots, cannot equal Captain Judikha of the Vortex Patrol. The idea of non-organic life is a theme in science fiction that goes back to at least the 1930s. The forms this concept has taken have ranged from intelligent types of crystal to planets inhabited entirely by self-replicating robots.

ABOVE: The Borg of *Star Trek* derived their name from "cyborg", which in turn is short for "cybernetic organism". This term, coined in 1960 by Manfred Clynes and Nathan S Kline, refers to any creature that is a combination of biological organism and machine.

OPPOSITE: James Cameron's *Avatar* (2009) was unique in not only depicting an alien race, but also in attempting to create an entire planetary ecosystem that was both scientifically consistent and clearly non-terrestrial. This environment, which was carefully developed in collaboration with several scientists, dictated the types of creatures that evolved on the world Pandora and the culture of its intelligent inhabitants.

for Manza, when the crystals are disturbed, the pattern is lost and cannot be recreated. The astronauts of *12 to the Moon* (1960) also encountered disembodied intelligences that communicated telepathically. Disenchanted with what they learn of Earth, they threaten to destroy it.

The Borg were introduced in the television series *Star Trek: The Next Generation* and played a pivotal role in the motion picture *Star Trek: First Contact* (1996). The Borg is a collection of different species who have been turned into half-machine, half-organic creatures that function as part of a hive mind called the Collective. These cybernetically enhanced beings can "assimilate" or absorb other races into the Collective in an unending effort to become a perfect race.

In *2001: A Space Odyssey* (1968), an unseen race of aliens direct human evolution over a period of tens of thousands of years. In *Interstellar* (2014), the unseen, seemingly omnipotent and omniscient aliens prove to be our future selves.

Overall, Hollywood aliens have evolved a great deal since the cultured, articulate and thoroughly humanoid Klaatu. Until the past decade or two, this reluctance to depict truly alien aliens was – understandably – driven by economics. With the advent of CGI, directors and designers are no longer limited to what is physically possible. This means not only more artistic freedom, but also the freedom to create aliens that really look as though they are the products of billions of years of evolution in an alien environment, so extending and developing the popular alien mythology.

A GALLERY OF SCIENCE FICTION ALIENS

Aliens depicted by science fiction illustrators have been as diverse as the artists themselves, their creations ranging widely over form and characteristics. Over the past century many, if not most, have been content with the standard humanoid or some variation on that theme. However, others have followed the lead of science fiction author Stanley Weinbaum and tried to depict aliens that genuinely appeared to be the product of an alien environment.

"Yeah, he came back, but he's not a goblin.
He's a spaceman."

Elliott, *E.T. the Extra-Terrestrial* (1982)

OPPOSITE, BELOW LEFT: H R Van Dongen created a convincing alien to illustrate Hal Clement's story populated with plausible non-human aliens.

OPPOSITE, BELOW RIGHT: J B Priestley's *Snoggle* (1971) anticipated many of the ideas that appeared later in *E.T. the Extra-Terrestrial*, although its title alien was certainly far removed from the vaguely humanoid star of the film.

BELOW LEFT: Artist Robin Jacques depicts a Martian in John Keir Cross' *The Angry Planet* (1945) that is part animal, part plant and part fungus. The author worked out a detailed ecosystem for the planet and followed this closely in describing the types of creatures that would live there and how they would behave.

BELOW RIGHT: Elliot Dold's illustration depicts jellyfish-like creatures from Venus seeking humans as food in Nat Schachner's story, "The Saprophyte Men of Venus" (1936).

OPPOSITE: Artist Wayne Barlowe (1958–) specializes in creating convincing and genuinely alien creatures. This painting from his book *Expedition* (1990) depicts a deadly aerial predator called the "skewer".

TOP LEFT: The meticulous paintings of Stephen Hickman (1949–) have graced the covers of hundreds of science fiction and fantasy books. This painting was created for the cover of *The Reaches* by David Drake (2004).

BOTTOM LEFT: A C Valentine created this watercolour depiction of robotic "animals" living on Mars in 1933.

ABOVE: Contemporary illustrator Tom Miller plays a whimsical riff on Grant Wood's iconic painting "American Gothic" in this illustration, where the 19th-century Americans are pitched against alien invaders.

THE GREAT MYTHS

The belief that aliens from other worlds are making – and have been making throughout time – regular visits to our solar system and Earth in particular has generated its own mythologies. Two of the most pervasive of these are the conviction that aliens have a presence on Earth's Moon, on Mars and possibly on other worlds as well, and the idea that if aliens have not only been visiting our planet for centuries – if not millennia – then they may have also played an active role in human history.

Aliens on the Moon and Mars?

A persistent myth has been the presence of aliens on Mars and our own Moon. The evidence for this is almost invariably based on astronomers and others trying to make sense of features teetering on the very brink of visibility. They then fall prey to a phenomenon that is hardwired into the human brain: *pareidolia*. This is the illusion of looking at something that is vague or obscure and thinking one is seeing something clear and distinct. The vaguer and more random the input is, the more likely it is that our brains will try to make sense of it, to find familiar patterns. As Carl Sagan pointed out in *The Demon-Haunted World*: "As soon as the infant can see, it recognizes faces, and we now know that this skill is hardwired in our brains." We've all done this consciously when trying to find shapes in clouds or in the cracks in a ceiling. There are entire pages on the Internet devoted to "faces" people have found in commonplace objects. The face of the man in the Moon is an example of pareidolia at work, as are the patterns of the constellations and the Elvises and Virgin Marys people have found on tortillas and burnt toast. When Percival Lowell observed Mars, he saw vague features at the very limit of the resolving power of his telescope – and his mind turned these into sharply defined canals crisscrossing the planet. It was probably the most notorious example of pareidolia until the "Face on Mars" was "discovered". Pareidolia may even account for a great many "flying saucer" sightings, in which the observer's mind fills in details that were never actually there in the first place.

The first person to report seeing artificial structures on the Moon was Baron Franz von Paula Gruithuisen (1774–1852). He was a Bavarian physician who later became a professor of astronomy and he made many important contributions to that science during his career (a lunar crater is named in his honour). Like many

LEFT: Interpreting complex patterns of light and shadow at the very limits of resolution, Baron Franz von Paula Gruithuisen (1774–1852) was convinced that he had discovered the ruins of an ancient civilization on the Moon. This was one of the drawings he created in which he tried to depict what he had seen.

others of his era, he believed that Earth's Moon was habitable – although most astronomers by that time were of the opinion that the Moon was all but airless and probably lifeless as well. However, Gruithuisen went a step further and was convinced that it was not only habitable but inhabited. This was something his predecessors Herschel, Schröter and others had also believed. He astonished his colleagues when he announced the discovery of an entire city near the Sinus Aestuum, north-northwest of the crater Schröter. He named this city Wallwerk for the pattern of linear ridges of which it was formed. They seemed to him to form a herringbone pattern that appeared to be artificial. He was certain that these ridges were in fact streets and buildings. He even identified a structure he called "Star Temple", which he suggested might be some vast monument. His announcement

caused a considerable stir in the press (and was undoubtedly the inspiration for the Great Moon Hoax of 1835). Even Tennyson was inspired to include a mention of Gruithuisen's lunar city in his poem "Timbuctoo":

I saw the Moon's white cities, and the opal width
Of her small, glowing lakes, her silver heights
Unvisited with dew of vagrant cloud.
And the unsounded, undescended depth of
Her black Hollows. Nay – the hum of men
Or other things talking in unknown tongue.
And notes of busy Life in distant worlds.

Perhaps seeing himself as a kind of new Columbus, a great celestial explorer, Gruithuisen presented his discovery not only to the public, but also to royalty and scientists. But then astronomers using much larger telescopes than Gruithuisen's tiny 6cm (2.4in) refractor started looking carefully at his Wallwerk and dismissed it

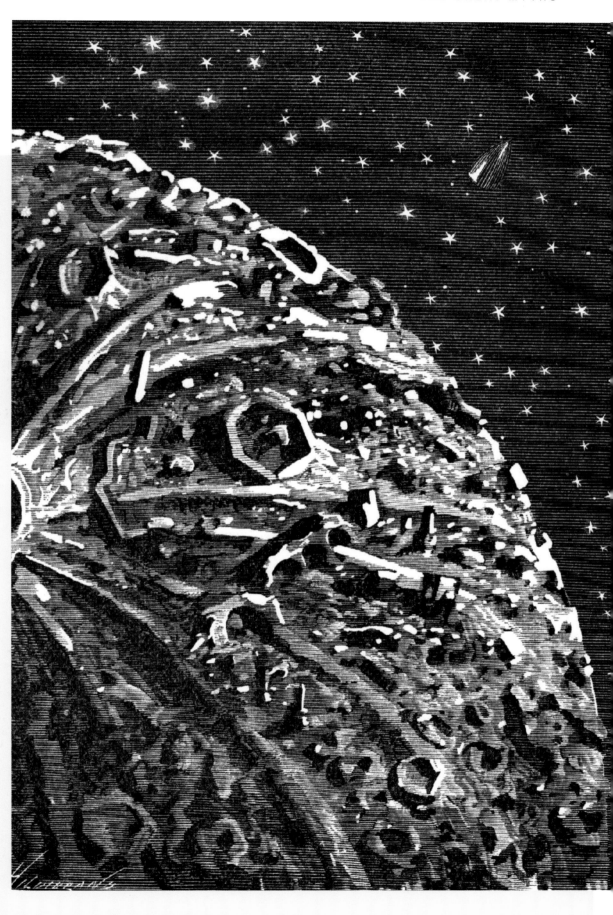

RIGHT: Jules Verne was probably inspired by Gruithuisen's discovery when the astronauts in *From the Earth to the Moon* (1865) spot what might be the ruins of an ancient lunar city as they orbit the Moon. This illustration is by Henri de Montaut from the original French edition.

as a natural landscape feature. Today, NASA spacecraft have imaged the area, showing that Gruithuisen's critics were absolutely correct.

Jules Verne's astronauts in his 1865 novel, *From the Earth to the Moon*, observed what appeared to be a ruined city as their orbit carried them above the surface of the Moon:

Michel Ardan . . . believed he recognized an agglomeration of ruins which he announced to the attention of Barbicane. . . . This stone pile, rather regularly laid out, appeared to be a vast fortress, dominating one of these long grooves which formerly were river beds in prehistoric times. . . . Below, he saw the dismantled ramparts of a city; here, the still intact curve of a gantry; there, two or three columns lying on their sides; further, a succession of arches which had once supported the conduits of an aqueduct; elsewhere, ploughed up pillars of a gigantic bridge, crossing the widest part of the groove. He distinguished all that, but with so much imagination in his eyes, through such a whimsical telescope, that it is necessary to defy his observation. But, which could affirm, who would dare state that the amiable lad did not really see what his two companions did not want to see? Moments were too valuable to sacrifice them to idle discussion. The Selenite city, alleged or not, had already disappeared into the distance.

For nearly a century there was not much serious attention paid to the possibility of the Moon

ABOVE: Astronomer John O'Neill announced in 1953 that he had observed an enormous bridge on the Moon. Although he did not state that it might be artificial, a great many other people jumped to that conclusion. It turned out, in the end, to have been an optical illusion. This recreation of what some people once thought the bridge might have looked like was created for an Italian book that was published in 1963.

ABOVE LEFT: The original NASA photo reveals the "iguana" to be a small part of a much larger image and just one more strange-looking rock among dozens of others.

ABOVE RIGHT: This Martian "iguana" is the result of too much imagination combined with low-resolution details extracted from a larger image. It is an enlargement from an original image taken by NASA's Curiosity rover.

being inhabited. The more astronomers were learning about the conditions on our natural satellite, the less likely it seemed that the Moon had ever harboured life. There was a flurry of excitement in 1953 when astronomer John O'Neill declared that he'd observed evidence for a huge bridge on the Moon. British astronomer Hubert Wilkins embraced the discovery with enthusiasm. "Its span", he wrote, "is about 20 miles from one side to the other, and it's probably at least 5,000 feet or so from the surface beneath." Wilkins was convinced that the bridge was the product of alien engineering and wasted little time in spreading the idea.

Needless to say, the fledgling UFO community seized on this idea with some considerable relish and developed it further. In the publication *The Flying Saucer Conspiracy*, Major Donald E Keyhoe declared it "an unbelievable engineering marvel apparently erected in weeks, perhaps in days." Keyhoe even excitedly hinted that there might be more to the story than was being revealed in the public domain, saying that "not even O'Neill dared to tell the whole story."

O'Neill himself thought the "bridge" was a natural feature and never claimed that it might be in any way artificial. As it turned out, the bridge was an optical illusion resulting from the complex shadows the rugged surface of the

Moon can create. Ironically, a real bridge was discovered on the Moon in 2010. The result of the collapse of a lava tube, it is not as impressive as O'Neill's massive structure would have been, but it is respectable nonetheless, with a span of 18m (60ft) and wide enough for a two-lane highway.

The beginning of the Space Age and the availability of close-up images of the lunar surface gave the old idea of aliens living on the Moon a shot in the arm. Armchair UFO hunters were soon scrutinizing every square inch of NASA and Earth-based telescopic photos, looking for even the slightest anomaly they could credit to an alien presence. It became a kind of cottage industry beginning with the work of author Joseph Goodavage, who, between 1974 and 1975, published a series of articles in *Saga* magazine "proving" the existence of an alien civilization on the Moon. Unfortunately, as space historian James Oberg points out, Goodavage was less than honest in his reporting, doing whatever was necessary to make his theory sound plausible and the government – especially NASA – sound as sinister as possible. As Oberg puts it, Goodavage "deliberately constructed a counterfeit mystery."

But he did his job well, setting the course for the hundreds of conspiracy theorists and UFOlogists who have carried his torch for the last 50 years. One outstanding disciple is Don Wilson, whose

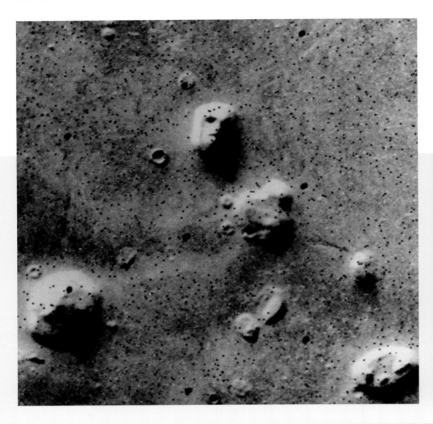

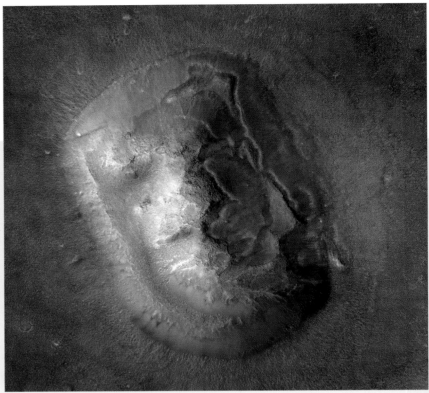

"We all confidently believe that there are at present . . . many worlds of life besides our own. . . . [This] may seem wild, and visionary; all I maintain is that it is not unscientific."

Sir William Thomson, later Lord Kelvin (1871)

bestselling books, *Our Mysterious Spaceship Moon* (1975) and *Secrets of Our Spaceship Moon* (1979), set out to prove that the Moon was not only a hotbed of alien activity, but was in fact a vast, hollow spacecraft in its own right!

Wilson, like many of those who have followed him, found evidence of alien "megastructures" on the Moon. These ranged from giant pyramids and colossal towers to enormous machines busily sculpting the lunar surface. Needless to say, all of these things were on the very verge of resolution and often required considerable squinting and imagination on the part of the viewer. Wilson was also frequently working with poor reproductions of the original photos. When first-generation images are examined, his "megastructures"

disappear entirely or resolve themselves into perfectly natural features.

There are a couple of outstanding problems with all of the aliens-on-the-Moon-and-Mars theories. The conspiracy theorists claim that NASA is actively covering up all evidence of an alien presence on those worlds. The problem is that they use easily available NASA images to prove the existence of alien artefacts. They seem to be completely unaware of the irony in this. The second problem lies, as we have seen, with pareidolia. Proof of this can be found in the fact that all of the perceived alien artefacts, structures and even animals that human beings believe they see on the Moon and Mars lie within barely discernible details, often on the very edge

ABOVE LEFT: The infamous "face on Mars"—which inspired a cottage industry of books and videos—in the end proved to be nothing more than an ordinary mesa (tableland feature in the terrain).

ABOVE RIGHT: This "dinosaur" skull found in another Curiosity image is a further example in the menagerie of creatures and objects supposedly found on Mars. It is also another example of pareidolia—the inbuilt tendency of the human brain to find meaningful patterns in random shapes. The original image showing the "skull" in context reveals it to be nothing more than an ordinary rock.

"A big greyish rounded bulk, the size, perhaps, of a bear, was rising slowly and painfully out of the cylinder."

H G Wells (1898)

RIGHT: "Bigfoot found stalking the surface of Mars" declared one tabloid headline in 2016. Like many such discoveries, the object is only a very tiny part of the original image and of tantalizingly low resolution. In fact, the rock is only a few inches high and only a few feet from the camera. The original panorama containing the figure of "Bigfoot" is enormous and filled with detail.

of human perception. One never sees clear and unambiguous photographic evidence.

As Sagan said, we are hardwired to find faces and the classic example of pareidolia at work is the infamous "Face on Mars". In 1976 NASA's Viking 1 spacecraft was photographing a region of Mars called Cydonia when it spotted the shadowy likeness of a human face nearly 2 miles from end to end. Scientists were startled, but quickly dismissed it as one of the many mesas (flat tablelands) that are common in the region. This one just happened to have some unusual shadows that made it look like vaguely like an Egyptian Pharaoh.

If the NASA scientists dismissed the "Face on Mars", the public certainly didn't. An entire conspiracy theory was evolved by people who were convinced that the face was absolute proof of life on Mars. On 5 April 1998 the Mars Global Surveyor flew over Cydonia for the first time. In

part due to public pressure and in part in the hope of ending the controversy forever, NASA had the spacecraft take a photo that was ten times sharper than the original Viking photos. Tens of thousands of anxiously watching conspiracy theorists and UFO buffs saw: a natural landform. There was no alien face after all.

But the theorists clung tenaciously. The photo had been taken through wispy clouds; perhaps the face had been obscured or distorted by the haze. So in 2001 NASA tried again. This time the team were able to use the camera's maximum resolution. What was clearly revealed, in unprecedented detail, was exactly what the scientists had expected all along: a butte or mesa not particularly different from those found by the hundreds in the American Southwest.

The face on Mars is by far the best-known alien "artefact" to be discovered, but it is hardly alone nor has it been the last. So far, enthusiasts have

FAR LEFT: The strange, disk-like clouds in Masolino da Panicale's painting *The Miracle of the Snow* (1428–32) have been interpreted by many as depicting a visitation from a fleet of alien spacecraft.

LEFT: In Edison's *Conquest of Mars* (1898), Garrett Serviss suggested that Martians visiting Earth thousands of years ago were responsible for building the pyramids and the Sphinx.

"identified" on Mars everything from gorillas, iguanas, bones and jelly doughnuts to crabs, Victorian women, squirrels and Assyrian gods. Every week brings some new "discovery" and, like all of the rest, it will surely be something on the very edge of perception or at the limits of a camera's resolution. Probably, just like the "Face", these will transform into commonplace rocks, shadows and other features when seen under better conditions.

Ancient Aliens

The idea that beings from other planets may have visited Earth in the distant past – perhaps interfering with the evolution of life on our planet – is not a new one. A related and more specific theme is that of ancient civilizations on the Moon that may once have had a part in life on Earth or even human history.

The 17 June 1864 edition of the French newspaper *Le Pays* featured a letter from an American geologist who claimed to have discovered a large meteorite. When broken open, a hollow chamber was revealed containing the mummified remains of an alien being. The humanoid creature measured about 1.2m (4ft) tall, with a round, hairless head and two large eye sockets. The arms were very long with five-fingered hands. Accompanying the body was a metal plate with engraved diagrams, similar to the disk that would be attached to the Voyager spacecraft in 1977. Decoding this revealed that the creature had arrived from the planet Mars some "millions of years" in the distant past.

The story was soon revealed to be a hoax, but not before many other newspapers – both in Europe and the United States – reprinted it as fact. It

ABOVE LEFT: The remains of an ancient lunar civilization, along with the skeletons of its giant builders, are discovered by the titular explorers of George Griffith's *A Honeymoon in Space* (1900). They speculate that as the Moon lost its air and water, its inhabitants were forced to live farther and farther underground – their bodies slowly, but unsuccessfully, adapting to the changing conditions.

ABOVE RIGHT: J A Mitchell also described a vast, ruined lunar city in his novel *Drowsy* (1917), which was dramatically illustrated by Angus MacDonall.

turned out to have been the brainchild of Henri de Parville, author of *Un Habitant de la Planète Mars*.

French science fiction author (and sometime collaborator with Jules Verne) André Laurie (a pseudonym for Paschal Grousset) had lunar explorers discover the remnants of an ancient civilization in his *Conquest of the Moon* (1889). So did the hero and heroine in George Griffith's *A Honeymoon in Space* (1901), who found pyramids and the fossils of a race of giants on the Moon. In John Ames Mitchell's *Drowsy* (1917), another space explorer also finds the remnants of a vast civilization that once existed on the Moon.

Garrett Serviss' *Edison's Conquest of Mars*, an unauthorized "sequel" to H G Wells' *The War of the Worlds*, went even further. He suggested that Martians invaded Earth 9,000 years ago, when they kidnapped humans, taking them back to

Mars as slaves. During their time on Earth, the Martians built the pyramids and the Sphinx, which it turns out was meant to be a monument to their leader. In addition to being one of the first to describe alien abductions, Serviss was probably also the first to suggest that aliens may have been responsible for creating ancient monuments, predating Erich von Däniken's *Chariots of the Gods* (1968) by more than half a century.

Although Otto Willi Gail's *Hans Hardt's Mondfahrt* (*By Rocket to the Moon*; 1928) doesn't contain any specifically extraterrestrial aliens, it does ring some familiar notes, such as when his space explorers discover a small asteroid on their way to the Moon. There, in a hermetically sealed chamber, they find the remnants of an altar. On a golden slab is engraved the image of a strange head: "a man's head, with long skull and

UFOs IN THE BIBLE

The Bible talks of a mysterious wheel that appeared in the sky, hovering over an astonished Ezekiel (Ezekiel 10:10). The description of what he saw sounds eerily familiar to modern readers:

… there were four wheels beside the cherubim, one beside each cherub, and the appearance of the wheels was like sparkling beryl. And as for their appearance, the four had the same likeness, as if a wheel were within a wheel. When they went, they went in any of their four directions without turning as they went, but in whatever direction the front wheel faced, the others followed without turning as they went.

Earlier in the same book (Ezekiel 1:4–5), the prophet – who seems to have been particularly inclined to seeing strange things in the sky – had another vision of celestial power:

As I looked, behold, a stormy wind came out of the north, and a great cloud, with brightness around it, and fire flashing forth continually, and in the midst of the fire, as it were gleaming metal. And from the midst of it came the likeness of four living creatures. And this was their appearance: they had a human likeness…

Looked at from a meteorological perspective, Ezekiel's wheel is not a little reminiscent of the halo phenomena that can occur when the sky is filled with high, icy clouds. These phenomena can be immensely spectacular, taking the form of concentric circles and radiating arms of light. To anyone not having seen anything like this before (and they may be rare in the Middle East), they would seem like something unearthly. Combined with a dollop of pareidolia and the sight can be

turned into almost anything. (The light from heaven that so impressed Saul in Acts 9:3 may have been a similar phenomenon.)

Zechariah wrote (in Zechariah 5:1–2) that: "Again I lifted my eyes and saw, and behold, a flying scroll! And he said to me, 'What do you see?' I answered, 'I see a flying scroll. Its length is twenty cubits, and its width ten cubits.'" This would have made whatever the prophet saw about 9m (30ft) long – reminiscent of the cigar-shaped UFOs that are often reported. And, of course, the endless descriptions of angels, elohim and nephilim descending from the heavens or semi-human nephilim living among humans are all fodder for the enthusiasts who would like to believe that the Bible is a literal record of alien visitors to Earth.

OPPOSITE: Ezekiel's biblical vision closely resembles the elaborate halo phenomena that can occur under certain conditions. It is easy to understand how impressive such a display could be to someone who may have never seen anything like this before.

RIGHT: Over the centuries, Ezekiel's apparition has been interpreted as everything from angels and extra-terrestrials to UFOs. In 1974 a NASA employee named Josef F Blumrich wrote an entire book attempting to prove that Ezekiel's vision was in fact the description of an extraterrestrial spaceship.

oblique, almond eyes." They speculate on what it could possibly be when one of the explorers notices what looks like an ankh symbol engraved on the forehead of the figure. "The primeval symbol of all earthly races!" shouts the explorer. Although they spend a few moments reflecting on what this means, the astronauts then continue on their journey to the Moon, leaving the reader to ponder what the significance of the strange figure might be until it is finally revealed when the remains of Atlantis are found on the Moon.

Although many other science fiction authors toyed with the idea of aliens visiting Earth in the distant past and perhaps interfering with biological or cultural evolution, the theme was pretty much relegated to the pulps until Erich von Däniken published *Chariots of the Gods* in 1968. Though the basic idea had been around for a long time, von Däniken presented the theory that aliens were responsible for not only most of the ancient civilizations on Earth, but their art, culture and monuments as well. He suggested that aliens not only visited Earth in ancient times,

ABOVE: Petroglyphs (rock carvings) appear all over the world and many depict strange creatures. A leap of the imagination has caused a large number of people to believe that they are evidence of ancient alien visitors. The petroglyph shown here is found in Val Camonica, Italy and may be more than 10,000 years old.

ABOVE: Nan Madol is a complex of massive stone structures constructed on a series of artificial islets in Micronesia. They were built between the 12th and 13th centuries, about the same time as the Cathedral of Notre Dame in Paris. However, many ancient alien proponents have credited them to alien interference, instead of attributing them rightly to the impressive engineering skills of non-European civilizations.

but were directly or indirectly responsible for monuments ranging from Stonehenge and the pyramids to Pumapunku and Nan Madol, the latter two being ancient stone complexes in Bolivia and Micronesia, respectively.

The theory has been questioned on several grounds. First is the assumption that thousands of years ago human beings were not as intelligent as they are now. Von Däniken's books have been based on the premise that ancient humans were incapable of developing their own art, technology, science and culture – they had to have these things brought to them by aliens.

The other problem is that most of the megalithic structures that are assumed to have required extraterrestrial assistance in their construction tend to be associated with non-white or

non-European peoples. Stonehenge is one of the few European examples, otherwise the position is that the ancient peoples of North and South America, the Pacific, Asia, Africa and the Middle East were simply incapable of conceiving and building large, complex structures. A counterargument is that the cathedrals at Chartres or Notre Dame had no assistance from extraterrestrial engineers, so it is not necessary to state that Peruvians needed alien help in constructing Machu Picchu, which was built at roughly the same time as these great cathedrals of Europe. Ever since the first of von Däniken's books appeared, professional archaeologists have argued against many of the ideas therein, and they will doubtless generate debate for years to come.

"... we are looking upon the result of the work of some sort of intelligent beings ..."

Percival Lowell (1894)

The Vedas – Sanskrit sacred texts written between 500 and 1500 BC – are often cited as containing evidence of ancient alien visitors, including descriptions of gigantic disk-shaped flying machines called "vimanas" that were made of metal and flown by uniformed pilots. Some readers interpret the texts as suggesting that the vimanas were powered by nuclear energy and carried incredibly powerful weapons. There *were* things called "vimanas" in Hindu mythology, but they were nothing like the giant flying fortresses described in UFO literature. The word *vimana* originally referred to palaces. Because temples are the palaces of the gods, *vimana* was used to describe the homes of the Hindu gods. Eventually, the giant flying chariots of the gods – drawn by flying horses or even geese – were also called vimanas. These chariots were depicted as flying on their own and later as being of enormous, even palatial size – though never losing the wheels that betrayed their origin. They were also described as having golden staircases, decorations made of costly gems, plastered terraces and even gardens filled with fruit trees. All very un-flying saucer-like, indeed.

The description of the flying saucer-like vimana comes not from legitimate ancient Hindu texts, but rather from a book written in 1952 by G R Josyer, *Vaimānika Sātra* (*The Science of Aeronautics*). This was allegedly dictated from the spirit world in 1918, but the first mention of it anywhere is in Josyer's book, where he describes the classic vimanas as resembling modern aircraft, with advanced propulsion systems and technology, that are even capable of flying from planet to planet – all things the legitimate ancient texts never claimed.

Other, more serious (if equally mistaken) researchers have suggested that primitive peoples may have cultural memories of alien visitors. In *The Sirius Mystery* (1976), Robert Temple proposed that the Dogon people of western Africa had been in contact with amphibious extraterrestrials about 5,000 years ago. Coming from a planet orbiting the star Sirius, the aliens – called the "Nommon" – left the Dogon with knowledge about Sirius that Temple believes they could not have possibly discovered any other way. However, Carl Sagan, among others, showed that the Dogon could have easily obtained this knowledge from European visitors, who had contact with the Dogon as early as the turn of the last century. Since Sirius is the brightest star in the sky and not surprisingly an important part of Dogon culture, it wouldn't be shocking to learn that discussions about their mythology would have soon turned to that subject.

Another popular ancient alien proponent who is taken seriously by believers is Zecharia Sitchin. His theory, expressed in a long series of books, is that aliens from the trans-Neptunian planet Niburu created humans by engineering the genes of female primates some 450,000 years ago. This theory has become the basis for a new faith called the "Raëlian Religion". Created by "Raël", an ex-racing car driver named Claude Vorilhon, it is based on the Sitchinian idea that humans are the result of a deliberate DNA experiment performed on pre-humans by aliens. The cult now has 50,000 members in 85 countries.

Is there any concrete basis to the idea of alien visitors coming to Earth hundreds or even thousands of years ago? Obviously, many people think so (a Google search for "ancient aliens"

TOP & BOTTOM: Several researchers have interpreted this Dogon drawing [top] as showing the star Sirius (X) and the orbit of its invisible companion (oval). However, the original drawing (bottom) reveals the presence of many different figures that had been selectively removed in order to prove the thesis.

OPPOSITE: Some passages from the Vedas, a sacred Sanskrit text, have been interpreted as describing enormous, nuclear-powered spacecraft called "vimanas".

returns more than 4 million results), even though there is no real evidence to support that belief. Still, that lack of proof has proved to be no impediment to the countless people who believe that every unusual historic event or artefact can only be explained by the interference of extraterrestrials.

Roswell

One of the most persistent of the modern myths – and the one perhaps most responsible for the perpetuation of the ubiquitous "grey alien" –

is that of the flying saucer that crashed near Roswell, New Mexico in July 1947. A crash from which it is claimed that alien bodies were recovered and analyzed.

One day a rancher named "Mac" Brazel discovered that some strange debris had fallen onto nearby Foster ranch. Being of an inquisitive mindset, and since he had been recently reading about flying saucers, he reported the find to the local sheriff, who, in turn, passed the information on to the commander of the nearby Roswell Army Airfield. The army sent investigators and what

ABOVE: Zecharia Sitchin (1920–2010) promoted the existence of a trans-Neptunian planet called "Niburu". Many of his followers believe that this planet is on an eventual collision course with Earth. Even a near-miss would be enough to cause worldwide devastation, as envisaged in this artist's interpretation.

"As far as I know, an alien spacecraft did not crash in Roswell, New Mexico, in 1947. . . . If the United States Air Force did recover alien bodies, they didn't tell me about it either, and I want to know."

President Bill Clinton, replying to a letter from a child asking about the Roswell Incident (1995)

ABOVE: When the US Army sent investigators to look into the reported crash of a "flying saucer" near Roswell, New Mexico in 1947, all they found were the remnants of a balloon designed to detect Soviet nuclear tests.

they found was not very spaceship-like, being only about 2.3kg (5lb) of aluminum and thin foil. In reality, they recognized it immediately for what it was: the remnants of a Project Mogul balloon. These consisted of long trains of balloons carrying ultra-low frequency antennae designed to detect Soviet nuclear tests. Needless to say, the project was ostensibly secret, and the authorities did not want it jeopardized.

Unfortunately for them, however, Brazel had contacted the local newspaper before calling the sheriff. The newspaper duly reported Brazel's conclusion that he'd found the remains of a

crashed flying saucer. The next day the newspaper printed a retraction, saying that the debris was merely that of a "weather balloon", along with Brazel's regrets for having spoken about the matter so hastily.

And that's where the incident stood until more than 30 years later, when UFO proponents, learning of the old, nearly forgotten incident, began mining the hazy and often conflicting memories of the few remaining living witnesses. Events of that single day were also conflated with unrelated events taking place over more than a decade. For example, a witness reported

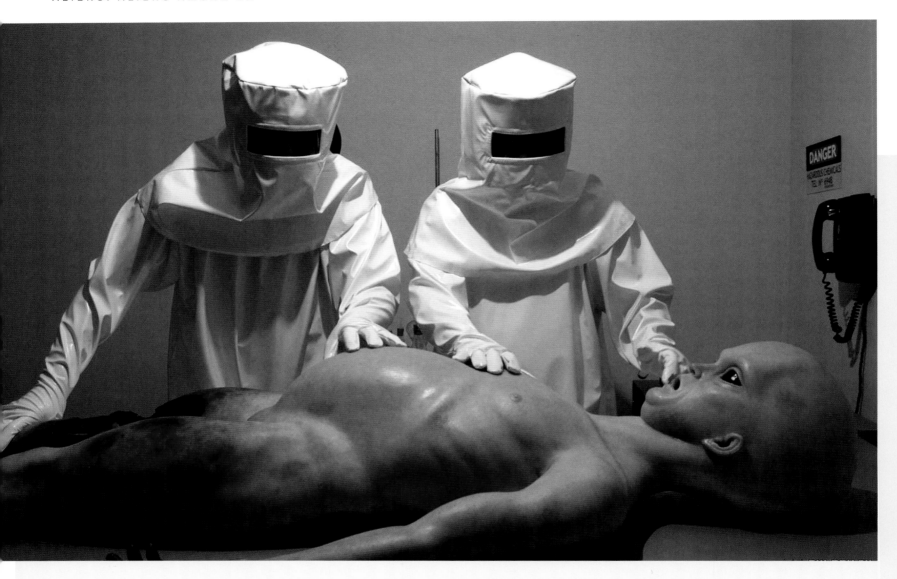

having seen an alien with a huge head walking into the air base hospital. Part of this is true: in 1959 an officer, who had been struck on the head by a balloon gondola, had his head and face swollen to the size of a beach ball, but was still able to walk into the medical centre under his own power. Another report of a witness having seen an autopsy performed on three small bodies that were blackened and mangled is consistent with the severely burned bodies of three airmen who had been killed in a crash near Roswell and who were autopsied there.

These and a dozen other pieces of similar "evidence" have grown exponentially over the decades until they have become a full-fledged myth and a cottage industry for the small town of Roswell.

And as the UFO myth has grown, it has become increasingly difficult to separate what may or may not be of legitimate interest from what is merely self-aggrandizement and patent commercialism. After all, books, movies and documentaries purporting to have evidence of the extraterrestrial origin of UFOs outsell their sceptical equivalents by many times. All this is understandable, for who likes a party pooper? There is certainly nothing wrong in having an open mind – a die-hard UFO fan who will accept no criticism is not so different from the die-hard sceptic who will accept no evidence – but there is a world of difference between having an open mind and accepting without question everything one hears or reads.

ABOVE: The infamous "alien autopsy" (here reenacted for a museum) was a film created in 1995 that purported to show one of the aliens recovered from the Roswell crash being dissected by a team of military doctors. The film was quickly exposed as a hoax.

OPPOSITE: UFOs and aliens have become a major industry for the small city of Roswell, New Mexico. As the city itself says, "The 1947 Roswell Incident is what put Roswell on the map and our UFO attractions, events and shops do not disappoint".

"I know for a fact the first UFOs reported in modern times, just before the crash at Roswell, were boomerang shaped and were reported as 'flying saucers' to describe the motion of their flight, like a saucer skipping over water. Yet immediately after, people saw and photographed saucer-shaped objects. Boomerang-shaped objects were rarely seen. Now people mostly report seeing large triangles instead of disks or boomerangs, because that is what they are told to expect to see."

Thomm Quackenbush (2013)

THE ALIEN INDUSTRY

That UFOs and aliens have been the basis of a thriving, and lucrative, industry for more than 75 years is a fact made evident nowhere better than in Roswell, New Mexico, where UFO tourism brings in $1.5 million annually. The International UFO Museum and Research Center is the imaginative focus of the attractions here, inspired by the infamous 1947 "crash" of an alleged flying saucer nearby.

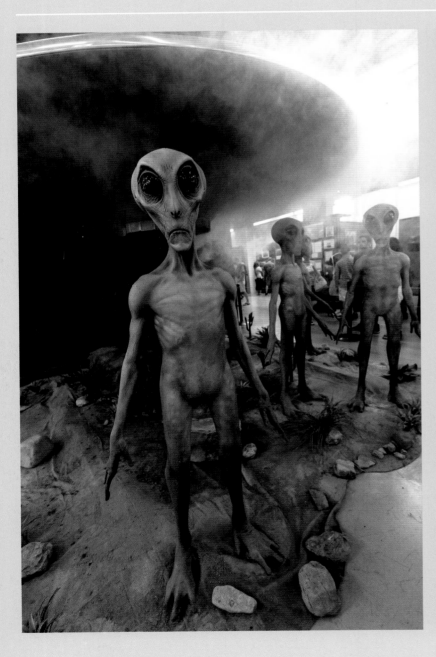

THIS PAGE & OPPOSITE: In addition to exhibits that promote the "grey alien" (opposite, left), the museum sponsors events, such as costume contests (opposite, right), lectures, films and art displays. It is also the hub of the annual Roswell UFO Festival. But the museum is not the only attraction on the "Extraterrestrial Highway" (officially State Route 375). Alien and UFO souvenirs are abundantly available and many businesses do whatever they can to attract their share of curious tourists.

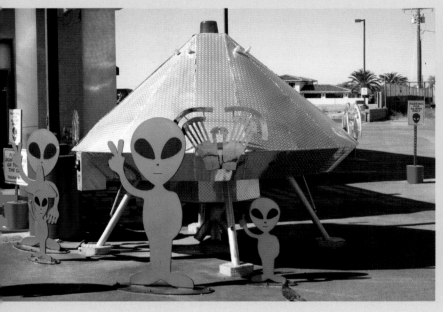

EPILOGUE

FIRST CONTACT: WHAT MIGHT ACTUALLY HAPPEN?

> "But what if SETI succeeds? Will we, as a sentient species, be changed forever philosophically?"
>
> Michael Carroll (2016)

OPPOSITE: Kelly Freas' illustration for the classic Frederic Brown novel *Martians Go Home* (1954) seems to depict a Martian's bemused reflection on humanity's obsession with extraterrestrials.

It almost seems naïve to expect worldwide panic at the news of the discovery of extraterrestrial life. Most people now simply take the possibility of alien life as a given. One poll found that as many as two-thirds of all Americans believe that aliens exist – a fact that is hardly surprising after more than a century of science fiction stories and movies as well as decades of scientific discoveries, such as the existence of thousands of other planets orbiting other stars. There is even an entire sub-genre of science fiction devoted to stories about "First Contact", which was the sub-title of a *Star Trek* film. For this reason scenes in the 1996 movies *Independence Day* and *Mars Attacks!* have crowds enthusiastically (if a little prematurely as it turns out) welcoming the appearance of alien visitors.

Possibly for these reasons, these same polls have shown that of those who believe in the possibility of aliens, most said they would be "excited and hopeful" if the existence of extraterrestrial life were to be confirmed.

Of those polled, 80 per cent thought that intelligent aliens would be more advanced biologically and technologically than humans while a full 90 per cent of those polled said that humans should reply to any message Earth might receive from outer space.

These ideas seem to be independent of political views, and are shared by conservatives and liberals alike. Perhaps not surprisingly, more than half of all atheists polled believe that alien life exists, compared to only about one-third of all Christians. The divisions between religious faiths is more striking. The religions of the world would face the most difficult problems with the discovery of extraterrestrial life, although each faith and denomination has its own particular view on the possibility. Buddhism already embraces a universe consisting of many thousands of inhabited worlds. Seventh-day Adventists and Roman Catholics are among the most welcoming to the idea. As early as the 1920s, the Abbé Théophile Moreux, founder of the Observatory of Bourges, argued for the existence

of intelligent life on Mars, and even wrote a book devoted to promoting the idea.

About the same proportion of Jews and Hindus as Christians believe in extraterrestrials, while more than 40 per cent of Muslims accept the possibility. Perhaps not surprisingly, Mormonism and Scientology are among the only religions that fully embrace the idea of extraterrestrial life. The key debate within the Christian church has been whether or not aliens would suffer from original sin and need to be saved by Christ. In 2014 Pope Francis suggested that any aliens, should they ever feel the need to visit Earth, would be welcome to be baptized.

Some believe that first contact with aliens might indeed have religious overtones. The first aliens to land on Earth may be missionaries come here to baptize humans in the name of their own god – a possibility that has not escaped science

fiction writers. Some of them have wondered if alien worlds have alien messiahs, while others have imagined Christ himself travelling from planet to planet. A famous novel by James Blish, *A Case of Conscience* (1958), explored the moral dilemma faced by a devout priest when a planet is discovered the inhabitants of which had never fallen from grace.

The stance of most religions regarding aliens would seem to echo that of many Jewish scholars, who suggest that there would be no problems so long as the personal relationship between God and the human race remains unchanged. It appears to be only the most fundamentalist followers of Christianity and Islam who reject even the possibility of extraterrestrial life. They adhere to the principle that not only Earth but the entire universe was created solely for the benefit of humankind.

ABOVE: Probably fuelled by Cold War fears, UFOs and aliens during the 1950s were largely associated with hostile invasions of Earth. Here the scene is a trio of flying saucers hovering over Washington, DC in the aptly named movie, *Earth vs the Flying Saucers* (1956).

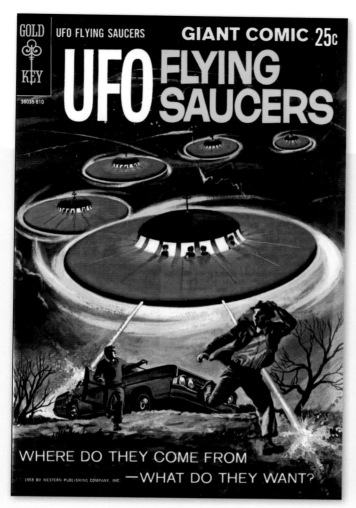

ABOVE LEFT: The questions asked by this 1968 comic book *UFO Flying Saucers*, "Where do they come from – what do they want?", might be answered by, "We don't know… and to take over the Earth."

ABOVE RIGHT: The cover of this 1954 pulp, created by artist Mel Hunter, suggests who might be the winner when it comes to human versus alien, especially if the conflict is on the alien's territory.

Other scientists tend to look at the consequences of alien contact a little more sanguinely. In 2016 the Philosophical Transactions of the Royal Society conducted a forum on the subject. Professor John Zarnecki and Dr Martin Dominik raised the issue of an international protocol regarding the meeting of aliens and humans and suggested that such a nonbinding protocol already exists in the UN's Committee on the Peaceful Uses of Outer Space. Created in 1959, the purpose of the committee is to facilitate matters of international concern within outer space. This includes the possibility of natural disasters, such as asteroid impacts, that would affect the world globally. Since communication between humans and aliens might likewise affect people around the world, an international political body such as the CPUOS would be ideally suited to oversee First Contact

and to control its impact on society. By the same token, Zarnecki and Dominik doubt that worldwide fear and panic would result from First Contact, at least not so long as there is no readily apparent risk – such as flying saucers vaporizing New York with lasers. They believe that the panic scenario is nothing more than part of the traditional alien contact myths. But there is no real reason to think this would happen.

But once First Contact is made, what then? We may have humans and aliens standing face to face (if the aliens stand and if they have faces), but what do they say to one another? Here enters the topic of modern "xenolinguistics" – how do we communicate with an extra-terrestrial race? Can we communicate? A handful of Hollywood movies have touched on this subject: *Close Encounters* (1977), *Contact* (1997), *The Abyss* (1989) and, most recently, *Arrival* (2016).

"Our sun is one of 100 billion stars in our galaxy. Our galaxy is one of billions of galaxies populating the universe. It would be the height of presumption to think we are the only living things in this enormous immensity."

Wernher von Braun, quoted in *The New York Times* (29 April 1960)

Like many of the scientists who have studied the problem, these films assume that communication between humans and aliens would be more or less a matter of decoding. Rather as if a Norwegian were to meet a Tahitian and neither spoke the other's language, it would not take long for two intelligent human beings to begin exchanging ideas and thoughts.

Yet it may be very likely that an alien race could be so alien that communication is not possible at all. To paraphrase Carl Sagan, a sufficiently alien race may not be able to tell the difference between a human being and a chrysanthemum.

Some of the scientists in the CPUOS forum thought that mankind should prepare for a worst-case scenario, preparing for First Contact in the same way they might prepare for an asteroid impact. Professor Simon Conway Morris thought that the evolutionary history of life on Earth – especially the unrelenting drive for survival of a species at all costs – might be duplicated elsewhere in the universe. Moreover, that instinct for survival might be what drives a species into space in the first place and to the exploration – and perhaps exploitation – of other worlds such as ours, with its natural resources. While it would appear that most human beings would welcome the discovery of intelligent alien life, and feel no threat to either their faiths or wellbeing, there are those, like Professor Morris, who urge caution. They point to the history of our own planet whenever a more "advanced" culture has come into contact with one that is less advanced. The classic (but by no means isolated) example is that of the Spanish discovery of the Incan empire in the mid-16th century. The native peoples were enslaved and forced to adopt the Christian religion. Those who did not were put to death. Foreign diseases decimated the population, the Incas' cities were destroyed and replaced by new ones built by the Spanish conquerors, and their language and culture were systematically extinguished.

H G Wells went even further and suggested that aliens might be so indifferent to humans that they would simply eradicate us as a matter of course, just as humans might eradicate an island of an inconvenient vermin before settling there. They may be so advanced as to not even distinguish us from algae and bacteria.

So when one is looking forward to the first contact between aliens and humans, one might be wise to stop and think, very carefully, about what one is wishing for.

RIGHT: Aliens as saviours or conquerors? This illustration, created in 1950 by Virgil Finlay, one of the grand masters of science fiction art, appears to offer an ambiguous answer to the question, the image both mesmerizing but also strangely threatening.

GLOSSARY

Accretion

The accumulation of particles into a massive object by gravitationally attracting more matter.

Amino acid

A large molecule that is one of the building blocks of life.

Anthropocene

The current geological age, during which human activity has had a major impact on the Earth and its environment.

AU (Astronomical Unit)

The mean distance of the Earth from the Sun, approximately 150 million km (93 million miles).

Biosphere

That region of a planet – encompassing the surface, atmosphere and oceans – within which life can exist.

Comet

An ice-rich interplanetary body. When it enters the inner solar system and is heated by the Sun, gasses are released that form a long, luminous tail.

DNA (deoxyribonucleic acid)

An extremely long macromolecule that is the main component of chromosomes and is the material that transfers genetic information.

Doppler effect

The apparent change in the frequency of light caused by an object moving toward or away from the observer.

Dyson Sphere

A hypothetical megastructure first proposed by physicist Freeman Dyson that would completely surround a star and capture most or all of its radiated energy.

Europa

The sixth of Jupiter's moons in distance from the planet and the smallest of the four known as the "Galilean satellites" in honour of Galileo Galilei, who first observed them.

Enceladus

The sixth largest moon of Saturn, the tiny (310km / 500-mile) satellite is a prime object for study for the possibility of life due to the ocean of warm water that exists beneath its icy shell.

Extrasolar hypothesis

The theory that UFOs are spacecraft originating on other planets.

Exoplanet

A planet orbiting a star other than the Sun.

Extraterrestrial

A living creature from somewhere other than Earth.

Extremophile

A living organism adapted to surviving in extremely hostile conditions.

Fermi Paradox

Essentially, the question: "Where is everybody?" Given the age of the universe and the number of potentially habitable worlds, the universe should be teeming with life, with evidence of it everywhere. However, this is apparently not the case.

Ganymede

The largest of Jupiter's moons and the largest moon in the solar system.

Geocentrism

The belief that Earth lies at the centre of our solar system – and perhaps also at the centre of the universe.

Goldilocks zone

The habitable region surrounding a star where a planet is not so close to the star as to be too hot for life nor so far away as to be too cold.

Great Dying

The Permian–Triassic extinction event that occurred on Earth about 252 million years ago, during which approximately nine in ten marine species and seven in ten land species vanished. Many causes have been suggested for this catastrophic event, including worldwide volcanism, an asteroid impact and even the explosion of a nearby supernova.

Greenhouse effect

The process in which a planet's atmosphere retains the heat the planet receives from its star, raising the surface temperature of the planet to a level greater than it would have without an atmosphere.

Interstellar

The space between stars.

Light year

The distance light travels in one year. At about 300,000km/sec (186,000 miles per second) that comes to approximately 9.5 trillion km (6 trillion miles).

Magnetotactic bacteria

Bacteria containing minute crystals that enable them to align themselves with the Earth's magnetic field.

Meteor

A rapidly moving streak of light visible for a few seconds in the night sky, , caused by the passage of a meteoroid or similar body through Earth's atmosphere

Meteoroid

A relatively small particle of rock or metal in space.

Meteorite

A meteoroid found on the surface of the Earth.

Moon

The name for the Earth's natural satellite. Often used casually to mean any small, solid body orbiting a planet or asteroid instead of the Sun.

NASA

National Aeronautics and Space Administration: the American civil space agency.

Nebula

A large cloud of gas and dust in outer space, made visible by reflected starlight or as a dark silhouette seen against brighter clouds or star fields.

Organic molecule

A molecule that contains carbon.

Panspermia

The idea that the basic "seeds" of life, such as complex organic molecules, can be carried from planet to planet by the pressure of stellar radiation.

Pareidolia

The inclination of the human brain to find meaningful shapes in randomness. This is especially manifested in a strong tendency to find faces and human figures in what are actually random patterns.

Planet

A solid or partially liquid body orbiting a star but too small to generate enough energy by nuclear reactions.

Planetesimal

One of the tiny, solid bodies orbiting the Sun that eventually coalesced to form the planets.

Planetoid

Also *asteroid*. A small, solid body or small planet orbiting a star.

Protoplanet

A collection of matter, mostly gas and dust, from which a planet is formed.

Protostar

An early stage in the formation of a star, when it is still gathering mass from its parent cloud of gas and dust.

Proxima Centauri

A dim red dwarf star that is part of a triple star system that includes two Sun-like stars. Proxima Centauri is also the star closest to the Earth (other than the Sun). An Earth-sized planet, Proxima Centauri b, was recently discovered orbiting it.

Pulsar

A rapidly rotating neutron star that emits beams of energy, which are detected on Earth as pulses.

The *Rare Earth* Hypothesis

The idea, made popular by Peter D Ward and Donald Brownlee, that circumstances allowing for the development of complex life forms are so extraordinary and so dependent upon chance that life in the universe would be a very rare phenomenon.

SETI (The Search for Extraterrestrial Intelligence)

An organization founded in 1984 to coordinate the efforts to discover the presence of extraterrestrial intelligence by monitoring radio telescope records.

Solar System

The Sun along with all the bodies – planets, moons, asteroids, comets, etc – that orbit around it.

Star

A mass of gaseous material large enough to have triggered nuclear reactions in its core.

Sun

The star orbited by the Earth and its fellow planets.

Spectrograph

An instrument that breaks light into its component colours. By examining the light from stars and planets with the spectrograph, astronomers can determine the elements of which they are made.

Stratosphere

The layer of Earth's atmosphere above the troposphere (the region immediately above the surface of Earth in which most weather takes place). It extends from about 6–10km (3.7–6.2 miles) to about 50km (32 miles) above Earth's physical surface.

Supernova

The explosion of a star, probably caused by gravitational collapse.

UFO

Unidentified Flying Object.

FURTHER READING

LIFE IN THE UNIVERSE

Barlowe, Wayne, *Barlowe's Guide to Extraterrestrials*, Workman, New York, 1979

Bennett, Jeffrey O, and Seth Shostak *Life in the Universe*, Pearson, New York, 2011

Chambers, John, and Jacqueline Mitton, *From Dust to Life*, Johns Hopkins University Press, Baltimore, 2013

Cohen, Jack, and Ian Stewart, *Evolving the Alien*, Ebury Press, London, 2002

Coustenis, Athena, and Thérèse Encrenaz, *Life Beyond Earth*, Cambridge University Press, Cambridge, 2013

Darling, David, *Life Everywhere*, Basic Books, New York, 2002

Dartnell, Lewis, *Life in the Universe*, Rosen Classroom, New York, 2011

Hartmann, William K, and Ron Miller, *The History of Earth*, Workman, New York, 1991

Lemonick, Michael D, *Mirror Earth*, Bloomsbury USA, New York, 2013

Lincoln, Don, *Alien Universe*, Johns Hopkins University Press, Baltimore, 2013

McAleer, Neil, *Arthur C. Clarke: The Authorized Biography*, McGraw-Hill, New York, 1992

Neumann, Megan, and Steph Minns, *Attack! of the B-Movie Monsters*, Grinning Skull Press, Bridgewater, MA, 2016

Scientific American (ed), "Exoplanets", *Scientific American*, New York, 2015

Time (ed), "The Search for Life in the Universe", *Time* magazine, New York, 2015

Ward, Peter, and Donald E Brownlee, *Rare Earth*, Copernicus, New York, 2000

Willis, John, *All These Worlds Are Yours: The Scientific Search for Alien Life*, Princeton, Yale University Press, 2016

ALIENS AND UFOs

Berlitz, Charles, and William L Moore, *The Roswell Incident*, Grosset & Dunlap, New York, 1980

Clark, Jerome, *The UFO Encyclopedia, Volume 1: UFOs in the 1980s*, Apogee Books, Detroit, MI, 1990

The UFO Encyclopedia, Volume 2: The Emergence of a Phenomenon: UFOs from the Beginning through 1959, Omnigraphics, Detroit, MI, 1992

Condon, Edward U, and Daniel S Gillmor (ed), *Final Report of the Scientific Study of Unidentified Flying Objects* (conducted by the University of Colorado under contract to the United States Air Force), Bantam Books, New York, 1968

Curran, Douglas, *In Advance of the Landing: Folk Concepts of Outer Space*, Abbeville Press, New York, 1985

Eberhart, George M, *UFOs and the Extraterrestrial Contact Movement: A Bibliography*, Scarecrow Press, Metuchen, NJ, 1986

Hopkins, Budd, *Missing Time: A Documented Study of UFO Abductions*, Richard Marek, New York, 1981

Hynek, J Allen, *The UFO Experience: A Scientific Inquiry*, Ballantine Books, New York, 1972

Jacobs, David Michael, *The UFO Controversy in America*, Indiana University Press, Bloomington, IN, 1975

Ramsey, Scott, Suzanne Ramsey and Frank Thayer, *The Aztec UFO Incident*, New Page Books, Wayne, NJ, 2015

Ruppelt, Edward J, *The Report of Unidentified Flying Objects*, Doubleday, Garden City, NY, 1956

Sagan, Carl, and Thorton Page (eds), *UFOs —A Scientific Debate*, Norton, New York, 1972

Sheaffer, Robert, *Bad UFOs*, CreateSpace, 2016

Scully, Frank, *Behind the Flying Saucers*, Henry Holt, New York, 1950

Sitchin, Zecharia, *The 12th Planet*, Avon, New York, 1978

Story, Ronald (ed), *The Encyclopedia of UFOs*, Doubleday (Dolphin Books), Garden City, NY, 1980

Strieber, Whitley, *Communion: A True Story*, William Morrow, New York, 1987

Vallee, Jacques, *Passport to Magonia: From Folklore to Flying Saucers*, Henry Regnery, Chicago, 1969

INDEX

PICTURE CREDITS

The publisher would like to thank the following people and organizations. Every effort has been made to acknowledge the pictures. The publisher, however, welcomes any further information so that future editions may be updated.

Image page positions are indicated as follows: L = left; R = right; T = top; TL = top left; TR = top right; C = centre; CL = centre left; CR = centre right; B = bottom; BL = bottom left; BR = bottom right.

Cover: Paul Palmer-Edwards
Endpapers: Getty Images/Mint Images – Frans Lanting

All images by Ron Miller or from Ron Miller's collection, except the following:

Alamy/ABN Images: 211: BR.
Alamy/A.F. Archive: 154; 185; 208.
Alamy/Frank Fotos: 211: BL.
Alamy/Granger Historical Picture Archive: 139: CL.
Alamy/Robert Harding: 203.
Alamy/MARKA: 202.
Wayne Barlowe: 188.
Richard Bizley, FIAAA, www.bizleyart.com: 66.
Copyright 2016 Kelly Freas: 212.
Getty Images/Bettmann: 138: BL; 139: CL.
Getty Images/Mondadori Portfolio: 198: TL.
Getty Images/MyLoupe: 211: CL.
Courtesy of Peter Greenwood: 147.
Image by Joel Hagen: 89.
© 2012 Stephen Hickman mixed media/digital: 143.
© 2004 by Stephen Hickman: 189: TL.

With the permission of the Estate of Betty Hill: 127: TR; 132: CL.
iStockphoto/Vertyr: 2–3
iStockphoto/isaresheewin: 54
iStockphoto/Oorka: 2; 14; 50; 114.
Burgess Shale © Carel Pieter Brest van Kempen by arrangement with Ansada Licensing Group, LLC: 64–65.
"At Home With The Tsailerol" by Karl Kofoed: 177.
Library of Congress: 41: CL; 140.
Thomas O Miller/Atomicart: 189: R.
NASA: 7; 24; 68; 77; 78; 79: TL, TR; 81; 83; 84; 111; 195: TL, TR; 196: TL, TR; 197.
National Oceanic and Atmospheric Administration (NOAA): 200.
Images used with the acknowledgement of the Frank R. Paul Estate: 70: TL, TR.
© Ludek Pesek, heirs, 2016: 125: TL.
REX Shutterstock: 179.
REX Shutterstock/20th Century Fox: 167: TL.
Rex Shutterstock/Century Fox: 170.
REX Shutterstock/Columbia: 168.
REX Shutterstock/ddp USA: 209; 211: CR.
REX Shutterstock/Courtesy of Everett Collection: 165: TL.
REX Shutterstock/Moviestore Collection: 167: TR; 175; 178; 184.
REX Shutterstock/Paramount: 165: TR.
REX Shutterstock/© Paramount/Everett: 173.
REX Shutterstock/StudioCanal: 172.
REX Shutterstock/© Universal/Everett: 174.
REX Shutterstock/ZUMA: 210: BL, BR.
SETI@home, University of California, Berkeley: 100.
Wikipedia/Matteo De Stefano/MUSE: 64

ACKNOWLEDGEMENTS

The author would like to thank David Brin and Dr John Elliott for their forewords and advice, Tom Miller, Karl Kofoed, Wayne Barlowe, Joel Hagen, Stephen Hickman, Peter J. Greenwood and Laura Freas for the use of their images and Chris McNab and Susannah Jayes, my indefatigable and patient editor and picture researcher.

AUTHOR

Ron Miller is an award-winning illustrator and the bestselling author of more than 50 books, including the Hugo-nominated *The Grand Tour* (over 250,000 copies sold), *Cycles of Fire*, *In the Stream of Stars*, *The Art of Space* and *The History of Earth*. He is a contributing editor for *Air & Space/Smithsonian* magazine, a member of the International Academy of Astronautics, and a Fellow and past Trustee of the International Association of Astronomical Artists.

CONTRIBUTORS

David Brin is a scientist, *New York Times* bestselling author and tech-futurist. His novels include *Earth*, *The Postman* (filmed in 1997) and Hugo Award-winners *Startide Rising* and *The Uplift War*. A leading commentator and speaker on modern trends, his nonfiction book *The Transparent Society* won the Freedom of Speech Award of the American Library Association. Brin's recent novel *Existence* explores the ultimate question: Billions of planets are ripe for life. So where is everybody?

Dr John Elliott is a Reader in Intelligence Engineering at Leeds Beckett University. He is the co-founder of the UK SETI Research Network and a member of both the SETI Permanent Committee and Post Detection Task Force. Often sought by the media for his expert opinion, Elliott appears in scientific magazines such as *Scientific American*, *Air & Space* (*Smithsonian*), *BBC Focus*, *New Scientist*, and *All About Space*, where he is ranked as one of the "5 most important people in the hunt for life on other Planets".